T0323535

Collective Behavior in Systems Biology

Collective Behavior in Systems Biology

A Primer on Modeling Infrastructure

Assaf Steinschneider
Protein Sequencing and Synthesis Laboratory,
University of Illinois, Chicago, IL, United States

ACADEMIC PRESS

An imprint of Elsevier

Academic Press is an imprint of Elsevier
125 London Wall, London EC2Y 5AS, United Kingdom
525 B Street, Suite 1650, San Diego, CA 92101, United States
50 Hampshire Street, 5th Floor, Cambridge, MA 02139, United States
The Boulevard, Langford Lane, Kidlington, Oxford OX5 1GB, United Kingdom

Notices
Knowledge and best practice in this field are constantly changing. As new research and experience broaden
our understanding, changes in research methods, professional practices, or medical treatment may become
necessary.

Practitioners and researchers must always rely on their own experience and knowledge in evaluating and
using any information, methods, compounds, or experiments described herein. In using such information or
methods they should be mindful of their own safety and the safety of others, including parties for whom they
have a professional responsibility.

To the fullest extent of the law, neither the Publisher nor the authors, contributors, or editors, assume any
liability for any injury and/or damage to persons or property as a matter of products liability, negligence or
otherwise, or from any use or operation of any methods, products, instructions, or ideas contained in the
material herein.

British Library Cataloguing-in-Publication Data
A catalogue record for this book is available from the British Library

Library of Congress Cataloging-in-Publication Data
A catalog record for this book is available from the Library of Congress

ISBN: 978-0-12-817128-8

For Information on all Academic Press publications
visit our website at https://www.elsevier.com/books-and-journals

Publisher: Stacy Masucci
Acquisition Editor: Rafael E. Teixeira
Editorial Project Manager: Sandra Harron
Production Project Manager: Maria Bernard
Cover Designer: Miles Hitchen

Typeset by MPS Limited, Chennai, India

Working together
to grow libraries in
developing countries

www.elsevier.com • www.bookaid.org

Contents

Assaf Steinschneider in memoriam

Assaf Steinschneider was born in Jerusalem in 1934, the son of Jewish immigrants who fled Nazi Germany. After serving in the Israeli military and becoming a kibbutz member he went on to complete a degree in agriculture at the Hebrew University of Jerusalem, Rehovoth campus and worked as an agricultural extension officer for the Israeli government. He came to the University of California in Los Angeles in 1959, where he studied horticultural science and moved on to the University of California at Berkeley to complete a PhD in molecular biology. He did postdoctoral work at UC Berkeley, taught at Albert Einstein College of Medicine, served on the faculty at the Hebrew University of Jerusalem, became a Research Associate at Medical College of Wisconsin, and ultimately moved to the University of Illinois at Chicago where he helped establish and head the protein sequencing/synthesis laboratory. Throughout his career Dr. Assaf Steinschneider conducted independent and collaborative research.

Assaf, our friend, was a serious cellist and pianist and a lover of classical music. In addition to a substantial library of science and math texts, he owned books on music history and musicology, musical scores, and hundreds of CDs and LPs. He was able to hold his own—even "hold court"—in discussions with musicians and scholars. He traveled extensively, familiarizing himself with the language, politics, history, and culture of each place.

After retirement, Assaf spent winters in his home in Jerusalem and the rest of the year in Chicago. During this period he began a broader study of systems biology and mathematics. He was passionate about his work, reading and attending conferences across the subfields of biology, physics, and chemistry. In the making for over a decade, we got weekly updates at our Friday night Sabbath table leading to the birth of *Collective Behavior in Systems Biology: A Primer on Modelling Infrastructure*.

Assaf Steinschneider died several months prior to publication of this book.

Stuart Kiken and Linda Forst

Introduction

A biological cell is a community of functional units: coding and catalytic nucleotide sequences, enzymes, signal receptors, and transducing proteins; the building blocks of transcription complexes, the apoptotic or antigen-presenting subcellular machineries, mitochondria and chloroplasts; up to and including normal or aberrant whole cells and teams, thereof. Evidently cellular biology depends on *which* of the building blocks are in the makeup and *what* they do. Much of what is already known originates in identifying cellular constituents, their structure, interactions, and biological roles. Generating predominantly *qualitative* information, this major approach to cell research now enjoys the capabilities of ever more powerful high-throughput technologies combined with constantly evolving traditional research methodologies. Already a significant source for new insights into cellular biology, new opportunities are following in medicine, agriculture, and biotechnology.

Given the makeup of its components, cellular life and the associated functions largely center on the *activities* of the functional units and the *processes* they carry out. Cells succeed when processes take place at a proper *level* of activity and at the right *time*, be it routine maintenance or in more specialized circumstances such as cell-fate decisions, adaptations to a changing environment, or response to various human interventions. Information on these largely *quantitative* aspects of cellular life, traditionally lagging, is now also becoming increasingly available due to recent methodologies.

How would the functional units, the cellular components, work together in a process? How does the *collective behavior* of the *whole* relate to the performance of the *parts*? The parts to the whole? A perennial question in science and elsewhere in human thought; how does it play out in cell biology? Aiming to provide answers by using *integrative* approaches that combine qualitative with quantitative features is an objective of *systems biology*. An ever-renascent and vigorously developing discipline, prominent early forerunners dealt with whole organisms and their populations.[1] Current emphasis is on cellular and lower-level genomic, proteomic, metabolic, and signaling *systems*.

---

1. Among early examples: (a) Thompson D'AW. On growth and form. Cambridge: Cambridge University Press; 1917, ibid, revised 1942, Mineola: Dover; 1992. (b) Lotka AJ. Elements of physical biology. Baltimore: Williams and Wilkins; 1925, Mineola: Dover; 1956. (c) Hinshelwood CN. The chemical kinetics of the bacterial cell. Oxford: The Clarendon Press; 1945.

Recent advances in studying these and other systems are becoming increasingly significant and thus of growing interest to the general cell biology community and to students entering the field—perhaps because they wish to keep up with new developments, but certainly to engage actively in realizing new opportunities.

Cellular processes as we know them typically enlist systems of multiple cellular components. They differ in quantitative and time behaviors depending on their nature and the cellular settings, that is, in a biological *context*. The ways to know quantitative and time behaviors of cellular processes can also vary and generally relate to the biological context as well. There is here a role also for scientific context—interpreting and assigning laws to cellular process behavior draws substantially on wide-ranging experience with natural and artificial systems elsewhere and the associated theory.

Getting to know a process's quantitative and time properties relies on a host of devices and schemes for describing and interpreting their behavior. What is currently available of these makes for a rich *infrastructure* that has meaning in its own right. The detail of working with the infrastructure is generally advanced, technical, and better left to specialists. However, a basic but useful appreciation of the infrastructure is within reach of the entire cell biology community—that is, an awareness of different available process descriptions and the infrastructure they employ; why, when, and how a particular infrastructure is employed for a given process; what information is needed as input to obtain meaningful output; what will be learnt about a process, possibly also about the host cell, perhaps more generally on how quantitative makeup can relate to biological function. There is, however, currently no literature that will allow common cell biologists access to this subject matter on their own. Intended as a primer for self-study, this book thus offers an elementary level survey of major current and emerging quantitative and abstract approaches to cellular system processes.

Quantitative methods for dealing with cellular processes at the molecular and immediate higher levels, at present the main focus in systems biology, are largely those familiar from chemistry and physics. To contend with heterogeneous and complex systems there is rich technological experience with parallels and a long-standing interaction with all levels of biology. How control and its design achieve particular goals essential in cellular life is one of its central themes. Dealing with higher-level cellular systems often turns to the concept of network, integrating quantitative behavior with information on system architecture. Whether as yet unknown guiding principles not necessarily vitalistic are still lurking within the complexity of biological cells remains to be seen.

Approaching cellular behavior in these terms calls for mathematics and related abstractions, or *modeling*[2] as it is known. A practice many in the

---

2. Using mathematical models as will be discussed in Chapter 1, Change/differential equations.

target audience are uneasy about, modeling does indeed raise legitimate concerns. Abstract descriptions, *models*, especially of complex systems, could misrepresent reality, being intentionally simplified, possibly omitting essential detail. At times models rely on particular assumptions, possibly hidden, that may not always be justified. One is often unclear whether and how quantitative theory relates to the experimental and real worlds. Personal discomfort with mathematics—its technicalities and its different culture—clearly also play a role. And then much of the success in cell biology so far has been based on qualitative approaches, traditionally guided by experience, common sense, and intuitive logic.

Still, keeping potential pitfalls in mind, mathematics and its abstractions are indispensable in fully exploring cellular processes. Derived from extensive real-world experience and backed by rich theory, mathematics is the standard format in science for describing quantitative and timed behaviors, well equipped to accommodate the diverse contributions of diverse cellular components, heterogeneous that they are. Mathematics also offers powerful methods for interpreting and enlarging on observed behavior: access to features not apparent otherwise, yardsticks for evaluating the choices cells make, criteria to test mechanistic models, and opportunities to generate ideas for new experiments, among other benefits, all the while using and generating well-defined, precise, and manageable information. However, mathematics does have a life of its own; it will have to be accepted by cell biologists as a full-fledged partner, as it is in other sciences and their applications.

Introductions to the methods in this book are plenty elsewhere, aiming for users who are oriented toward mathematics, the physical sciences, and engineering and their common level of preparation. The primary background expected here will be in basic molecular cell biology and related disciplines. Previous exposure to elementary algebra, calculus, and probability theory will be helpful, but working knowledge is not assumed. Whatever is needed from these disciplines will be provided in the text; readers will be responsible for no more than simple arithmetic.

Whether infrastructure or context, the selection of materials for this book naturally reflects the author's background as well as a judgment of what readers may want to, or perhaps should, know. Underlying is experience that began with higher-organism farming, centered on a professional lifetime in molecular life sciences research, and more recently draws on the learning experience associated with writing this book. Elementary and informal, this work is not intended to be exhaustive or balanced, but mainly to illustrate a variety of major and promising approaches. Generally starting each topic with basics, there will be an effort to rationalize the underlying motivation, real world and mathematical. Relevance to cell biology is a high priority. Also covered will be technological sources, important context that is unlikely to be accessible to this audience otherwise. Interspersed there will be a few more detailed examples, highlighting how infrastructure can be applied in specific cases. Fully realistic examples would generally be advanced and detailed beyond the present scope. Altogether, mathematical technicalities will be kept to a minimum; the few awe-inspiring symbols will be demystified where needed.

Turning to the infrastructure, two themes, here related, will provide a setting. *Change*—quantitative and its management—is the first and main focus of this book. Often of the essence in biological function, it is a common defining feature of processes in cellular biology and elsewhere in the real world. Dealing with change the main emphasis will be on *dynamics*, the way change occurs, up to the life span of one or a few cell generations. *Organizational structure*, meaning *functional relationships* between cellular constituents and their larger *patterns* of action and interaction is a major topic in cell biology in its own right. A largely permanent feature, it has direct bearing on cellular dynamics and their control and in that context will here be the second theme.

Prevalent in quantitative handling of cellular processes, the approach predominant in this book will be *deterministic*. Familiar from course work, time-tested, and instructive, it presumes that essential features of collective behavior can be captured with high *certainty* and that the future states of ongoing processes are *predictable*. Much of the emphasis will be on *linear* approaches. Mainstays of process modeling with generality and power, they are also relatively straightforward in introducing basic concepts at an entry level. Methods for dealing with common nonlinear behavior, realistically, will have their place, but most of that detail will be left to more specialized sources.

To model process dynamics, *differential equations* are commonly the first choice in cell biology as they are elsewhere in the sciences and their applications. Taken up first, this will be the basis for what follows. Characterizing systems with large and heterogeneous membership may employ *linear algebra*. Practical, alternative representations that substantially extend the scope of the methodology are available in *approximation* and *numerical* methods. Providing insights into higher-level systems, well-characterized or only partially known, *systems* and *control* theories will emphasize the technological point of view. Quantitative approaches to *optimality* will tie in with biological success. For the sake of completion and departing from the deterministic approach, there will be a short introduction to the dynamics of cellular processes that are better understood in terms of what is *probable* as based on *random* or *statistical* properties, an area of increasing interest. Turning full force to organizational structure, presenting complex cellular systems as *networks* will conclude the book. This introduces *graphs* and theory that draw on both the mathematical and the visual experience and is supplemented by *logical* devices that are useful in linking organizational detail with network function. Taken together, the emphasis will be on infrastructure that deals with process behavior directly—that in a model which comes closest to biological significance and presumably is the main interest of the target audience.

Contemporary modeling practice has become sophisticated to the point where creating and managing models are supported by theory of their own. Devising approaches to data shortage and optimizing large computations are important. To assure model quality it may be necessary to evaluate whether models are sufficiently representative, accurate, robust, computable, or adequate by other criteria. Weighty considerations all, to obtain an adequate account readers will have to consult more specialized literature elsewhere.

Computers play a major role in modeling cellular biology and much else in the real world. However, aiming for generality, process models are commonly formulated, first in terms of mathematics and related abstractions such as those in this book. Machine computation will be employed later on, depending on the case. Computers work with mathematical approaches that differ considerably from that of the initial model. The two can be bridged, but the transition and associated computer technicalities can be significant factors in originally choosing, formulating, and handling models. These topics will be touched upon here but the computer end, that is, hardware and software, will not.

The subject matter in this book is arranged in a natural sequence and the main points in each chapter are presented mostly in the text. The book's contents are not intended to be mastered in all their detail. Readers are encouraged to be flexible about what to follow, in what order, and how much detail to aim for. One could sample first from the topics more obviously oriented to real-life systems, control, and networks and turn to the mathematical detail at a later time. Whichever the approach, it will take motivation, none will be effortless. A second pass and more are recommended. A few more advanced references will be listed including original papers and URLs, accessible online. Author's errors, inevitable or not, will hopefully be forgiven.

All said, this book is intended, first and foremost, to create in lay cell biologists an awareness, or perhaps an initial orientation to the infrastructure for modeling collective behavior in cellular systems. Readers interested in more advanced literature,[3,4] will find here an opening. Those already

---

3. As good starting points: (a) Aleksandrov AD, Kolmogorov AN, Lavret'ev MA, editors. Mathematics: its content, methods, and meaning. Moscow: Russian Academy of Science press; 1956, English: Cambridge, MIT press; 1963, Mineola: Dover; 1999. Accessible background; examples largely from physics and engineering. (b) Morrison F. The art of modelling dynamic systems. New York: Wiley-Interscience; 1991, Mineola: Dover; 2008. An eminently readable practical perspective. (c) Edelstein-Keshet L. Mathematical models in biology. New York: McGraw-Hill; 1988, Philadelphia: SIAM; 2005, a more demanding standard in its field.

4. Recent textbooks include: (a) DiStefano J. Dynamic systems biology modeling and simulation. Amsterdam: Elsevier/Academic Press; 2014. Comprehensive, with engineering perspective. (b) Klipp E, Liebermeister W, Wierling C, Kowald A, Lehrach H, Herwig R. Systems biology. Weinheim: Wiley-VCH Verlag; 2009. (c) Helms V. Principles of computational cell biology. Weinheim: Wiley-VCH Verlag; 2008. (d) Philips R, Konev J, Theriot J, Garcia HG. Physical biology of the cell. New York: Garland Science; 2013. All are highly instructive, but assume some familiarity with mathematics introduced here later on.

exposed to this content elsewhere will gain a different perspective. Individuals contemplating active research will hopefully recognize in what follows some pointers toward new discoveries.

Acknowledgements: I am grateful to several individuals that provided input during the preparation of the manuscript but bear no responsibility for its final version. Especially Drs. S. Agmon, Z. Marx, and the late A.J. Yahav. I am also highly indebted to Drs. S. Kiken and L. Forst for moral support in various forms. And thanks to Rafael Teixeira, Sandra Harron and Maria Bernard for their careful attention to all details and for seeing this through to publication.

A. Steinschneider

Chapter 1

Change/differential equations

Context

A biological cell alive. A community of functional units in its *collective behavior*. That is, a community of receptors, enzymes, molecular complexes in gene replication and expression, the cytoskeleton and molecular motors, organelles, the machineries of antigen presentation and apoptosis, among others. *How much* and *when* they are active is obviously critical to the workings of whichever the cell and whatever the functional units. Whether it is a particular cellular function or a whole life cycle, cellular biology seen this way becomes a matter of the *quantitative* and *time* behaviors of cellular *systems*[1] and their *processes*. Among the latter are cellular housekeeping, genetic and epigenetic control, response to environmental cues, the generation and management of stress, the mechanics of cellular and tissue morphogenesis, and electricity-based information transfer by nervous conduction, to mention a few.

Take, for example, the process of eukaryotic cell division. High drama under the microscope, it has been extensively described in images and words. But the complete picture of how mitosis actually works will require quantitative data, that is, on the chemistry and structure of the DNA and its associated proteins and their interactions, as well as those of the building blocks of the mitotic spindle, the positions of the moving chromosomes and the underlying forces, the curvature and physical qualities of emerging nuclear envelopes, and cell membranes or walls, among others. And, not in the least, on when processes began and ended and time spent in-between.

When investigating cellular quantitative and time behavior, *what* should be represented? *How* can it be done? What will it *teach*? Could there be *universal* ways to deal quantitatively with cellular processes, diverse and complex as they may be? Are there properties that would allow a description in the same terms, an enzyme reaction, intracellular substance transport, the

---

1. What a *system* is, is a matter of definition and depends on the context (see Chapter 4: Input into output/systems). Here and throughout this book the term, system, will refer to homogeneous or heterogeneous communities of cellular or other real-world functional units that perform a particular task collectively. System members do, or do not, interact directly within the system or with the surroundings.

Collective Behavior in Systems Biology. DOI: https://doi.org/10.1016/B978-0-12-817128-8.00001-8

bending of cellular membranes, or the actions associated with electric potentials? Perhaps the throughput of an ensemble of different enzymes in a metabolic pathway? Even the actions of complex heterogeneous processes such as the control of the eukaryotic cell cycle where multiple processes of a constitutive signaling machinery are exquisitely coordinated with induced syntheses, modifications, and interactions of a variety of nucleic acids and proteins? There is a common denominator to these and other real-world processes, quantitative *change*. Abstract, but readily observed, and here serving the purpose.

Quantitative change is obviously a characteristic of cellular biology at all levels—the dynamic state of cell constituents, intracellular transport, differentiation, circadian rhythms, senescence, and others. Ditto the lives of multicellular organisms, their populations, not in the least, Darwinian evolution. Initiating, sustaining, and controlling change are central to medicine, agriculture, and industry. Change is also a major theme in the overall scientific view of the real world and in its applications, as well as in scientific methodology. Recall that deciphering cellular processes from protein folding and enzyme action to metabolism, gene expression, and cell replication, up to tissue differentiation or response to the environment, among others, owe much to varying experimental conditions—time, reactant concentration, molecular environment, or genetic makeup, and so forth. Ubiquitous and diverse, descriptions of quantitative change can take various guises such as *kinetics, dynamics, process, behavior, development, evolution,*[2] among others. Not necessarily mutually exclusive, some of these terms may and will be used here interchangeably.

The point is that quantitative change is an opening to the use of a wide range of tools for representing and interpreting processes available in mathematics. Together these provide an *infrastructure* for characterizing change that is in plain view. Also, when it is not immediately apparent, notably at steady states in living cells and elsewhere, a major feature of real-world processes. Some of the methods extend to, and have a major role in, dealing with altogether permanent properties of matter at the molecular and atomic levels. Mathematics offers a direct and universal handle on process behavior in the real world, the main focus here, and does this in well-defined, precise, and efficient ways. It is also key to using machine power—computers are a major and often indispensable component of the methodology.

Whether in cell biology or its applications in medicine, agriculture, and various biotechnologies, mathematical tools are applicable to cellular processes at all levels of organization. An infrastructure for various purposes,

---

2. Everyday language can be appropriated differently in various disciplines. In case of doubt about a term, context and common sense should be helpful. Note that process *behavior* in this book will often refer to quantitative characteristics of change and the word *information* will be used strictly intuitively and not bound to any abstract or quantitative theory.

mathematics can apply to basic processes, the parts, as well as to process combinations that make for a collective whole. Moreover, its methods can serve to identify and characterize higher patterns in collective process behavior and its response to change in settings such as the internal cellular makeup or the environment. Time being of the essence in cellular life as well as in its study, the emphasis in this chapter will be on applications of quantitative methods to ongoing process behavior over time. Being versatile, the methodology also will be useful in later chapters for other purposes.

Now, aiming to rationalize cellular or other real-world behavior, especially in mathematical terms, it is sensible to turn first to those processes that display *regularities* and presumably do obey some law(s). Most cellular processes, as we know them and as represented in this and other chapters of this book, answer to this description and are *deterministic*. When monitored, systems presumably yield adequate information, change is taken to be regular and predictable, and observations are assumed to be representative and repeatable. These notions are not necessarily self-evident or philosophical niceties. *Random* behavior, inconsistent with these assumptions, is an alternative that is increasingly coming to the fore in our understanding of cellular processes. Each type of process has to be dealt with on its own special terms. Deterministic behavior will be taken up here, random behavior in a later chapter.[3]

Natural, experimental, and technological change is often *continuous* in appearance—even when in reality it represents the collective behavior of a random population of small, distinct units, for example, enzyme molecules catalyzing a biochemical reaction, virus particles replicating, microbial cells growing in a fermentation setup. Continuous processes such as these often obey surprisingly simple rules; their behavior finds a good match in *differential equation(s)* (*DEs*), the main theme in this chapter and ubiquitous elsewhere in this book. Backed by extensive theory and time-tested methodology, *DEs* have a central role in the sciences and in technological applications. In describing living cell processes at all levels, they offer major tools for dealing with cellular process *dynamics*—ongoing change of the moment as well as overall process properties. Behavior of the latter, especially of complex processes under modified conditions, is addressed by the *qualitative theory* of *DEs*.

---

3. *Deterministic* descriptions routinely apply to so-called *macroscopic* systems that obey the more familiar quantitative laws of classical chemistry and physics. Applicable to much in everyday science, these descriptions extend down to the molecular level, in these disciplines considered *microscopic*. This is so that an average value is adequately reliable, given that system members are present in sufficient numbers (minimally in the 100s). On the other hand, small number microscopic systems typically display *random* or *statistical* behavior. Such is evident and increasingly documented in single-cell processes involving single digit copy enzymes and transcription complexes. Calling for *probabilistic* descriptions, these will be discussed separately in Chapter 7, Chance encounters/random processes.

Change can be stepwise in appearance when it takes place in discrete entities that are identifiable and countable, such as gene copies, organelles, mitotic figures, and embryonal cells, among others. Seen as *discontinuous* or *discrete* processes, they can be represented by *(finite) difference equations*. Currently in limited usage with cellular processes and guided by logic that is often similar to that of working with continuous *DE*s, discussion here will focus on the basics.

Employing continuous and discontinuous equations of change often calls for methods from other mathematical disciplines. These will be introduced below, with additional explanations in later chapters. Complex and heterogeneous systems are thus often represented by multiple DEs and may call for methods of *linear algebra*. Mathematical limitations may be managed by turning to alternatives, usually *approximations*, extending also to *computer*-compatible *numerical methods* and *simulations*. Altogether, infrastructure in this chapter is common in quantitative approaches to cellular processes which emphasize lower level processes. It is also increasingly being used to deal with higher level systems, control, and networks and in conjunction with organizational features.

Assembling a statement

Collective behavior

In the quantitative approach taken here, cellular processes represent the collective behavior of *systems* of cellular components. What will this mean for the infrastructure in this chapter and later in this book? A process such as mitosis evidently represents collective behavior of multiple cellular components and associated processes. But what can be said about simple, elementary-level processes such as adrenaline binding to a receptor, glucose reacting directly with hemoglobin, or trypsin breaking a peptide bond? Almost all of what is known about these and other relatively simple interactions between cellular components is the outcome of collective behavior. Ditto higher level processes such as protein synthesis, signal transduction, genetic control, viral replication, among much else. This is so because, for one, it is in substantial numbers that observations on cellular components can actually be made—experimental and other, in vivo and in vitro. Reckon also that one is looking for regular behaviors to begin with, likely a collective property even in simple, relatively homogeneous systems (as pointed out in footnote 3). What follows in this book will then largely deal with collective behavior in processes, simple and complex, that take place in large-copy−number cellular components and that is suitable for the deterministic descriptions this allows for. That said, there will also be the customary exceptions. Cells also display collective behavior that is statistical, for example, the activities of single-cell, low-copy-number transcription machinery, the partial reactions in catalysis by single enzyme molecules, or that of the residues in proteins and nucleic acids that change conformation. A short detour into what is relatively recent and now a vigorous avenue of research will be taken in Chapter 7, Chance encounters/random processes.

Essentials

How will a quantitative statement be made? Used? Procedures for dealing with cellular or other real-world processes, deterministic or other, commonly begin with creating a mathematical process representation, a *model*. Facing the overwhelming detail of the real world, biological cells in particular, creating and working with models or *modeling* is the practical way to go. Models succeed by focusing on essential features while disregarding what presumably is subsidiary; this is by choosing a particular process feature as it relates to one or more known or assumed underlying factors, sometimes called *governing factors* or *decision variables*, among others. Typical in cellular modeling are, for example, models of metabolic and signaling pathways as they depend on the activities of their lower level members. Mining for new information commonly takes further mathematical workup.

Models can serve different purposes. Beginning with raw data, one might aim for a mathematical expression that fits the data, explains the observations, and opens access to other process characteristics. Based on knowledge already available, models can serve to test hypotheses about mechanism, recognize new features of process behavior, and generate ideas for further study. Sometimes, as with the familiar visual models, mathematical models are a convenient way to summarize existing knowledge about a system, possibly as a formula that expresses a rule that was indeed obtained by modeling.

What might quantitative process models consist of? Most commonly, models are formulas for change, equations whose terms represent factors of interest and how they (might) relate. Many cellular processes, for example, are often represented by the balance between synthesis and degradation of a particular cellular component say protein as seen in Eq. (1.2). When dealing with organizational features, there may also be visual graphs and other abstract devices, as will be seen in later chapters.

Offhand, the same cellular process can be represented by various models—there is no one-size-fits-all, standard formula. Making a choice will depend on the type of process, purpose, and information already available. That the mathematics is workable, and how, is a significant consideration, as well and a major theme in what follows here. As diverse as quantitative models of cellular behavior or other processes may be, there is a common format in which they are set up. Moreover, special rules may apply as to how that should go.

Model makeup

A cell experiences a considerable reduction in size when it divides. It may, in its lifetime, accumulate mutations. Both are examples of *absolute* change and can be of biological interest; however, how cellular factors relate, a major theme in system behavior, is commonly studied and expressed in terms of *relative* change. The finding that quantitative change in two factors is correlated will help establish cause and effect and presumably shed light on the underlying mechanism(s).

Variables

To represent what changes during a process and how, mathematical models typically describe relative change by two types of *variables*. One, the so-called *independent* variable, is chosen by the modeler to represent, say, a factor underlying a process—input into the model, if you will. Actual process behaviors that result, output, will be represented by *dependent* variables. Modeling, for example, glucose-dependent ATP synthesis in cellular respiration, the amount of glucose consumed by experimental design could be the independent variable, x, and the amount of ATP produced the dependent variable, y. Although not the only conceivable measure of linkage between variables and the factors they represent, most popular is relative change represented by the *ratio* between the respective changes that both undergo, the *rate* of change. Using Δ (delta) as a symbol for *total* change in either variable, the *quotient* $\Delta y/\Delta x$ is the elementary-level prototype. *Time* being an important factor in cellular life, as well as in observing cellular processes, models of change over time are ubiquitous, as they are in modeling other real-world processes. In those cases, *t(ime)* is the independent variable replacing x.

Relating the variables

Formally the relationship between independent and dependent variables is defined by the familiar mathematical concept of *function*. These days, mathematical functions are commonly presented in terms of set theory—mappings from a domain to a codomain, among others. Quite adequate and more useful throughout this book will be the older and more intuitive notion of a function—a relationship, or formula if you will, that specifies how dependent variables relate to variables that are independent. Employing symbols, for instance the number of bacterial mutants, y, arising as *f(unction)* of carcinogen added, x, will be denoted by $f(x)$ where x is the *argument* and $f(x)$ is the function (of x) that will provide a value of y corresponding to x. Other notations for the same abound in the literature. In real life, change is often due not only to one factor, as above, but to several. The observed concentration of cellular protein, for example, often depends on both synthesis and degradation. As seen in Eq. (1.2) the two contributing processes differ in quantitative behavior and each requires a special mathematical function of its own, represented by a separate term. Formally, the relationship between overall protein concentration and both partial processes taken together is also a defined mathematical function. Models in the literature and later in this book that are represented by simple statements, such as $y = f(x)$, in fact often consist of multiple terms, each having a separate function.

Choosing functions for a model, both their *existence* and *uniqueness* are perennial mathematical concerns. Neither is to be taken for granted. A function of x, or $f(x)$, is said to *exist* where it is able to assume a value. Obviously essential in representing the real world, it is not always realized mathematically. Since, for example, division by zero is impossible, functions that employ fractions such as $1/x$, or $1/(1 - x)$ do not exist at the points $x = 0$ or $x = 1$, respectively. Functions such as these can still be useful elsewhere provided they are well defined over the points of an appropriate *interval* of x (i.e., between two *end points*).

To set up a representation that is meaningful, member functions must be *unique* (i.e., allow for the one correct value, only). Problems arise in various ways. As a classical example, there is the circle, described by the equation $x^2 + y^2 = r^2$, where x and y are the coordinates and r, the radius. Since y satisfies the equation for every value of x equally well whether it is positive or negative, $f(x)$ is not unique. To avoid potential ambiguity, modeling with equations of change, including those here, one commonly chooses functions that are *single valued*. Being single valued is, in fact, ordinarily implied when using the term, function, as it will be here. That particular types of mathematical expressions actually exist and are uniquely applied often requires substantial proof, but will be taken for granted from here on. Finding out which functions would be appropriate, then, becomes a main goal.

Terms collectively

Given that each component of the makeup meets these requirements, there may be additional considerations in using them and putting them together in specific cases.

Format

A simple format, ubiquitous in the literature, is in a sense an initial declaration of intent, a provisional list of the factors of interest. Later on, it could be fleshed out with quantitative detail to make for the familiar equation format. As an example that will be revisited later on (see Eq. 1.4), there is a common model of the cellular throughput (or turnover) of, say, a metabolite such as a sugar or an amino acid, the so-called *flux*. In that model, flux is described by $f(X, t, D, x, y, z, v)$ (i.e., as function of the concentration of the substance, X, the time, t, its diffusion coefficient, D, the coordinates, x, y, z, and the rates of its various reactions, v).

To make a more definitive statement will take full-fledged terms that, not surprisingly, will be assembled into *equations*—defined over a particular interval when needed and, again, preferably single valued. An equation that represents the familiar *equality* ($=$) is arguably the most practical and satisfying. Also noteworthy here are the weaker "is approximately" ($\approx$ or $\cong$), and the "proportional to" ($\propto$). An *inequality*, larger than ($>$) or smaller than ($<$), or a mixed statement, equal to or more than ($\geq$), or equal to or less than ($\leq$) may represent, for example, certain threshold phenomena or purely mathematical circumstances (e.g., when a particular function only exists over part of the range of observation). It is not uncommon to have cellular and other real-life models that employ a mix of these and other relationships to represent different conditions of the same process.

Terms on either side of an equation collectively make for the *left-hand side* (LHS), or the *right-hand side* (RHS). Complex processes may call for grouping equations together in *systems of simultaneous equations*, an important theme later on. Recall that (as will also be assumed here) in representing the real world, both sides of an equation have to agree in their *physical dimensions* (e.g., time, mass, length, and volume). While to start with process model equations will include terms for all the factors considered relevant, one often legitimately drops terms when their contribution is negligible in particular circumstances, for example, dropping the second term in Eq. (1.2) when protein degradation is insignificant.

Linear (in)dependence

Representing a process, one would offhand expect complete freedom in selecting functions for a representation. Employing equations with multiple terms, however, may lead to multiple and, therefore, ambiguous outcomes, unless the original functions are *linearly independent*. Well definable mathematically, linear independence implies that different *terms* of an equation are not similar enough, say, to be interconvertible by simply changing or attaching numerical constants. In other words, an equation would not be allowed to include both $2x^2$ and $5x^2$ as separate terms because multiplying the first by 5/2 will yield the second. However, an equation with $5x^2$ and $5x$ terms which cannot be interconverted this way is acceptable.

Linearity and nonlinearity

Given the freedom allowed by linear independence, one would naturally aim for model components that best represent the nature of the process and the purposes of the model. This, not surprisingly, will determine the mathematics that can be used with the model and can affect how one would collect data as well. Naturally or by experimental design, some cellular processes take place at a constant rate, for example, the new synthesis of certain induced bacterial proteins or many enzyme reactions. In these cases, the increment in product stays constant over time and the amount of product is described by a straight line. Processes and behavior such as this are known as *linear*. Linearity is displayed when sums of contributions to a process are *additive*. If, for example, an operon codes for three proteins, say, in the amounts x, y, and z, and assuming degradation is negligible, measured newly synthesized materials will add up to $x + y + z$, a linear relationship. Additive, linear behavior correlates with the mathematical makeup of the variables. It is commonly associated with variables of the *first power*, say concentrations such as [X], [Y], [Z]. Being additive is implied also in multiplication by a constant, essentially shorthand for multiple additions. The equation $y = 2x$ accordingly assumes the values 2, 4, 6, ... when $x = 1, 2, 3, ...$; its curve is a straight line and the increment $\Delta y = 2$.

From the cooperative binding of oxygen to hemoglobin to the different phases of the eukaryotic cell cycle, and much else, *nonlinear* cellular processes are common, possibly having special biological roles. Product increments in nonlinear processes typically vary over time and lower level contributions are not additive. Nonlinear behavior often arises due to the mundane reality that cellular processes involve functional units that interact directly. For example, when a cellular receptor, I, dimerizes, there will be a term with $I \times I$, or I^2 (i.e., the contributions of the reactants will be represented by a *product*, not a sum as in the linear case). Described by the equation $y = x^2$, when $x = 1, 2, 3, 4, ...$ increments between values of y are 1, 4, 9, 16, ... (i.e., variable) and the equation is nonlinear, so are equations with *higher power* variables or with *mixed* terms. The latter typically arise in the ubiquitous cases of *different* factors interacting directly. For example, (neglecting other machinery) m-RNA, M, and a small ribosomal subunit, R, forming a complex, MR, could be represented by a product of the concentrations, [MR]. In more special cases nonlinear behavior arises because of a particular quantitative makeup of more complex equations. It can be the case when contributions made by a variable and a constant are comparable, a possible scenario, for example, with Michaelis−Menten enzyme reactions when the substrate concentration is comparable to the K_m.

Being linear or nonlinear also can relate to how processes are observed. Offhand, the ideal way to collect data would be right from the beginning and without interruption. In practice this may not be feasible. Under linear conditions it would not matter much, but the less-regular, nonlinear behavior can make fitting data with a quantitative model problematic to begin with.

As it were, not only describing data, but also the actual mathematics of nonlinear processes, an example of *nonlinear mathematics* is considerably more difficult to manage than are linear models. In favorable cases, nonlinear behavior can be handled directly by advanced formal or graphic methods.

A limited option in dealing with this ubiquitous issue, modeling cellular processes may go in different ways. For one, and a major option as will be seen, is to turn to less rigorous mathematical methods. However, it is often possible to describe nonlinear behavior with linear expressions adequately enough for the purposes of a model. It may be warranted by adjusting the experimental and other real-world conditions as suggested above, for example, by collecting data over a limited range of observations, such as very short times or in a specific range of cellular component concentrations. In some cases, nonlinear contributions are relatively small and can be ignored. Often more practical is to make models linear using mathematical arguments. There are in fact standardized *linearization* formulas that will be taken up later in this chapter and in greater detail in Chapter 3, Alternative infrastructure/series, numerical methods.

All said, whenever there is a choice in quantitative modeling, cellular processes or elsewhere in the real world, the *linear* format is favored. As it were, work with nonlinear models often turns to a linear format sooner or later, as well. Considering also the variety of its applications, linear modeling will thus play a major role throughout this chapter and elsewhere in this book.

A bird's eye view

Processes represented by an equation or a curve that depicts its trajectory are often evaluated from different points of view. Change and its mathematical properties are considered *local*, as seen at one particular point of a curve or in its immediate neighborhood. Properties are *regional* over longer intervals and *global* when they relate to curve and process in their entirety; as an example, a microbial culture and respectively its: (1) momentary biomass change any time, (2) mass increase over the log phase, or (3) total biomass accumulated over its entire history including steady state and decline. Common models originally describe extended time periods or an entire history of a process, discussed in more detail in Chapter 4, Input into output/systems. However, as will be seen, mining for information often depends on methods that are geared toward either local or global properties and, seldom, both.

Change by differential equations

An algebraic prelude

Everything considered, what might a quantitative process model end up like? To begin with a minimal version, consider once again the respiratory breakdown of glucose (glc), but now in aerobic glycolysis where one molecule of glucose theoretically will yield 38 molecules of ATP. A barebones representation of this process could read $[ATP] = C \cdot [glc]$ where the concentrations

[*ATP*] and [*glc*] are the dependent and independent variables, *y* and *x*. They are linked here by a (linear) function familiar in the form $y = f(x) = ax$, where *a* is a constant. Both *C* and *a* in this equation are *coefficients* of the respective independent variables. Both are constants that represent how the function would apply in specific case with *C* here maximally equaling 38. In terms of the particular system represented, *C* is also a *parameter*, a quantity that does not change during an experiment or other observation but may take another value under different conditions, say, modeling anaerobic glycolysis where *C* would equal 2.

Evidently these examples are algebraic expressions familiar from everyday life. With *y* on the LHS, terms on the RHS will be variants of the most general, also known as *polynomial*, form, ax^n. They may thus harbor a coefficient, *a*, a variable, *x*, and a positive exponent, *n* or $1/n$ when raising to a power or taking a root. When more terms are present they will be of the same form and interrelated by *arithmetic* operations, addition and subtraction, multiplication and division. Raising powers and taking roots of groups of terms is allowed too.

Algebraic equations starting with such as the minimal version above serve in modeling a wide range of static and dynamic features of the real world. Describing cellular and other processes, they will be useful later on in this and following chapters in various ways: in representing deterministic but discontinuous change, in providing a mathematically convenient format for alternative methods in work with difficult equations, in characterizing more permanent process features including patterns of ongoing process behavior, as well as the modeling of organizational structure and optimizing performance. Not in the least, as the mathematics employed in computer operations.

Turning differential

Modeling continuous cellular processes predominantly employ the *DE* (plural *DEs*). Cellular processes that can be represented by *DEs* ranging from the most elementary to entire pathways and up to those of whole cells and in the test tube, as well. *DEs* are also instrumental in characterizing other cellular process properties, such as stability to interference and other response to changing conditions. Among others, they find uses trying to identify optimal performance as will be seen in later chapters.

Turning to *DE* models generally carries a number of benefits: (1) *DEs* provide a convenient mathematical format for representing *ongoing* change; (2) expressions in *DE* infrastructure can describe a *wide range* of processes; and (3) there is extensive apparatus for *generalizing* from process behavior in a particular setting to how it might behave in other conditions as well as for other types of interpretation.

How might these factors play out? Ideally, modeling would lead to rules for process behavior over an entire spectrum of plausible conditions, be it time, temperature, concentration, and so forth. Such rules would preferably be rooted in a wide range of actual observations. In reality, collecting data is commonly feasible only for a single or a limited set of conditions. Progress is possible, nevertheless, based on the notion that change as seen in particular data reflects a more general process property, presumably related to an underlying mechanism. This property in turn determines also how the process will behave in other settings. The crux here is that the general law and its particular applications are defined so that they can be obtained from the particular data essentially using mathematics, only. For example, measuring the change in concentration or amount of newly synthesized protein for a limited time in a particular experiment will be the opening to specify the general law, a formula that allows determination of how much protein is in the system at any time. The formula can be used further to obtain additional formulas for protein levels in different times and settings.

For an intuitive angle on how modeling with *DE*s might work one can turn to parallels in less-formal cell biology. Appreciating how cells function as well as developing applications, such as effective medicines and pesticides or profitable fermentation processes, can involve two types of insight. One is an understanding of how cellular processes would work in general, say, particular mechanisms of action, which processes contribute to a particular biological function—how, and in what range of conditions they can take place. The other concerns how processes might behave in a specific case, say, the throughput of a metabolic process in certain normal or pathological cells. Modeling with *DE*s, as will be seen, deals with two analogous layers of information.

Modern *DE*s originated with Leibniz and Newton in the 17th century. The latter employed them, among others, to explain the motions of the planets, but without knowing the particular contributions of other stars; in other words, the rest of the system, a situation quite familiar in biological cells. The name, DE, is a historical carryover from the differential notation employed for the derivative at that time. *DE*s have since become major tools in many sciences and their applications, as well as in pure mathematics. A vast field with extensive literature, what follows in this chapter will focus on *DE*s likely in modeling cellular processes. For those interested, a short review of infrastructure for *DE* models is provided next.

Building blocks and operations

Functions

DE models obviously should employ mathematical functions—the expressions that relate the process to the chosen governing factors—that reflect quantitative and mechanistic properties of the real-world process. Models should also be open to mining for information by further mathematical workup and should provide actual process values when needed. Among the more common expressions of an offhand, much wider choice for the terms of *DE*s are the algebraic and the transcendental, such as e^x, log x, and sin x, real and complex. Extending the capabilities further are the additional calculus-based building blocks derived from these functions, applying the associated operations—*derivatives* and *derivatization*, and often *integrals* and *integration*. To relate the different governing factors, represented by equation building blocks, *DE*s still manage with the arithmetic operations above.

Modeling continuous processes naturally calls for functions that have the property of *continuity*. It obviously is essential if the model is to cover a continuous process everywhere. Continuity is also a prerequisite in much of *DE* mathematics. In that context continuity means that change is gradual, all points of an interval exist, and, except at the ends, all points have neighbors. Formally definable and rigorously testable, continuity will be taken for granted in this book where it applies: even only one missing point (i.e., a single break in a curve) could make a function non*differentiable* or non*integrable* by standard procedures. *Discontinuities*, such as these and others, nevertheless, occur in real-world processes and can also arise from mathematical operations. There is indeed a substantial effort to deal with discontinuities that will here be left to more specialized literature.

Limitations on choosing building blocks arise also when linear *DE* models are called for. Functions such as e^x or sin x, among many others, are inherently not linear and therefore excluded from initial models, although they can arise during workup later on. It turns out that linear models consisting of quite simple algebraic expressions can still represent a wide repertoire of cellular behaviors.

Moreover, choosing model components also takes into account what the form of the outcome will be (i.e., the solution to the problem). Modelers preferably turn to equation components whose workup will yield *elementary* functions. Here, these would be mainly algebraic expressions such as the above, the exponentials e^x or $(exp)x$, the logarithms $\log_e x$ or ln x (the natural logarithm), and trigonometric functions such as the sine, sin x, and the cosine, cos. Among these, especially noteworthy here are the *exponential* functions. They arise in modeling many familiar, natural, and artificial processes—deterministic as well as statistical. Typically, the rates of these processes depend on the total number of system members at a particular time point and on a constant proportion undergoing change—unrestricted growth of microbial cells, progress of chemical reactions, or radioactive decay, for example.

Derivatives

Creating a *DE* process model, say, of how the concentration of a metabolite depends on the processes by which it is synthesized and removed will require expressing change in terms of the variables and functions of the model. Once again, modeling turns to the *ratio* of the change—in process values, say, metabolite concentration, the (dependent) variable *y*, and in the contributing factors the (independent) variable(s), *x*. This ratio, $\Delta y/\Delta x$, is known as the *rate of change*. Using this ratio, but in the more refined form of the *derivative* of a function, describing a process is at the core of modeling change with *DE*s. Derivatives of functions here play a dual role. They serve to maximize the fidelity of representing change and they provide the mathematical basis for making generalizations, to begin with, by finding integrals. It will be useful to turn here, first, to some basics. An informal review of derivatives, related integrals, and coefficients is provided next. Readers already in the know may safely skip these, but they could probably find something new in the short comment on operators that follows.

How do derivatives describe change? To begin with one can take a process that simply follows a straight line, say, line A in Fig. 1.1. The rate of change, $\Delta y/\Delta x$, is constant and the same from the beginning to the end. Corresponding to an overall *average* rate of change, it can be calculated using simple arithmetic. Averaged change can be calculated similarly for other curves by using a straight line that connects both end points, such as line B between the origin x_0 and point x_a in Fig. 1.1. Evidently these are inaccurate; one could get a better rate of change by more closely mimicking a curve with shorter lines between nearer interior points. This is practical in some cases and likely inadequate in others. Especially problematic could be complex process curves rich in detail such as Fig. 1.2, say, a close look at the putative distance between two cellular organelles, or at the generation of protons to maintain cellular pH. Whichever is applied, to capture maximum

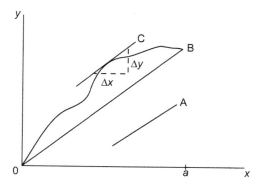

FIGURE 1.1 The derivative as a slope.

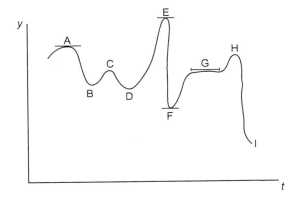

FIGURE 1.2 Stationary points.

detail of continuous processes, models would have to specify change essentially at every ongoing stage, at every point of their trajectory.

Change, as seen so far, implies a difference between at least two stages of a process that would be represented by two different points on a trajectory. How can change and its rate be dealt with at only a single point? A way to do this is offered by the mathematical concept of *limit*—intuitively meaning here that the value of the dependent variable, y, will tend to a particular value no matter how much the independent variable, x, changes. As applied to curves, limit change at a particular point would be represented by a straight line that might have connected two points, as in Fig. 1.1, but that now, as also shown there by line C, has been moved to touch the curve at that one point, only. That line is the familiar *tangent*; its rate of change, $\Delta y/\Delta x$, corresponds to its *slope*, and its value can be readily calculated using the point of contact together with any other second point. Turning from this geometrical description to one that employs variables and functions, or $\Delta f(x)/\Delta x$, a limit of y at a particular value of x can be obtained, imagining ever shorter increments of x. In an abstraction that carries the argument to a logical extreme this measure of change is indeed the derivative of the function that relates y and x, at point x. In formal terms the derivative is the mathematical limit of the total change ratio, $\Delta f(x)/\Delta x$, when both variables become infinitely small. Its value equals that of the slope of the tangent at that point.

Due to the originators, derivatives are commonly represented by dy/dx, a still popular historic *d(ifferential)* notation (Leibniz), and by the symbol y' (Newton). Obtaining a derivative from a function is an operation known as *derivatization* or *differentiation*. An operation that can apply differently depending on the parent function, detail is available in standard calculus textbooks. Once derivatized, a function may be open to additional derivatization. Taking *higher* derivatives, for example, of algebraic functions repeatedly n times, symbol $(d^n y/dx^n)$ follows the well-known formula y $(x^n)' = nx^{n-1}$. That a function is indeed *differentiable* n times will depend on its makeup. Using this formula, differentiation of $y = x^2$ is, for example, possible only twice, since third or higher order derivatives would not exist.

In addition to the typical usage of derivatives for local behaviors, process descriptions sometimes employ derivatives to specify larger scale behaviors, regional and global. When plotting, say, the distance between or the amount of a cellular component (y) versus the time (t), as in Fig. 1.2, and values of y increase with t at a particular point, as well as over a stretch or an entire curve, the first derivative is positive, $y' > 0$; and when y values decrease it is negative, $y' < 0$. It is, however, when first derivative equals *zero* (i.e., $y' = 0$) and the straight tangent lines in the figure are horizontal that one is dealing with special features of a process that, here, are of particular interest. For one, as will be recalled from calculus, they are indicative of *maxima* and *minima* in real-world properties. Such would be represented respectively by Fig. 1.2, points A, C, E, H, and B, D, F, I. Among those, point A has the highest value and therefore is a *global* maximum; the others are *local* maxima. Similarly, point I, with the lowest value, is the global minimum; the other minima are local. Using derivatives this way can establish process optima as will be seen in Chapter 6, Best choices/optimization. Zero valued first derivatives are also characteristic of processes at steady state—the stationary phase typical in many a biological and technological process in which it displays no net change, but has actual yet constant throughput. As will be seen later in this chapter and elsewhere in this book, modeling the steady state can be a major source of information on collective behavior, especially in complex processes. Turning to Fig. 1.2, depending on the particular model, a steady state would correspond to a saddle point such as G. The first derivative there indeed equals zero, as would the slope of the tangent.

Derivatives introduced so far, *ordinary* if you will, relate processes to a single independent variable, in cellular modeling often the time. Recall that the above barebones model of metabolite throughput, the Flux, also involved the coordinates in space, x, y, and z—three additional independent variables. This calls for *partial* derivatives. Taking change in each variable to be independent of the others, these will be handled like ordinary derivatives one at a time, keeping the others constant. In common notation, the differential symbol d is replaced by ∂, and flux with respect to time and the coordinates will be given by $\partial F/\partial t$, $\partial F/\partial x$, $\partial F/\partial y$, and $\partial F/\partial z$, respectively.

Before moving on, a more general comment about derivatives, ordinary and partial, is still in order. As calculated, both ordinary and partial derivatives are, strictly speaking, approximations, albeit close, of an idealized mathematical entity that already approximates real-world behavior. The point is that while some might expect "mathematical precision" in a model, employing derivatives is a first in a series of approximations that may be called for in modeling change in the real world, as will be seen later in this chapter and elsewhere in this book. Nonetheless, using derivatives of functions still makes available the powerful tools for dealing with and generalizing on the wide variety of quantitative relationships in the real world, now available in mathematics.

Integrals

Given the derivative(s) that describe(s) a process, the way is open to what was referred to, above, as a general law for change. With much more explanation later on, this means finding functions that would give rise to a given model if they were to be derivatized, a sort of mathematical backtracking, if you will. Called *antiderivatives*, these functions in principle are obtainable by *integrating* the model equation, the product of the operation being its *integral*. In essence this amounts to finding a function that represents the *sum* of the infinitely small, instantaneous, ongoing change, represented by a derivative. The symbol of taking an integral, $\int$ is in fact a stylized S(um)-shape.

In favorable cases integration will follow familiar textbook methods of calculus and will yield an *(indefinite) integral*, also a mathematical function. In the simplest case, the indefinite integral of x^n, or $\int x^n dx$, can be obtained simply by inspecting the derivatization formula above and working back to find that it is $x^{n+1}/(n+1)$. The indefinite integral is commonly accompanied by an unknown constant, C, which offhand does not say much about the nature of the process, but will be very useful later on.

Now, if the indefinite integral is a formula for what is actually changing, say, the actual concentration of a cellular substance, a major goal of modeling is to know exactly by how much, in other words, as expressed by a *numerical* value, a real *number*. The indefinite integral can serve this purpose in different ways if it is provided with additional process specifics. A major theme in this chapter and this book, establishing ongoing process values given their initial value, will be described at length later on.

Definite integral

Modeling many a real-world process aims for the total value it generates during particular stretches or over its entire lifetime, for instance, of the total biomass produced by a cellular culture or the profit generated by a commercial enterprise. To obtain that value, one would use an indefinite integral and specify between what limits the evaluation will take place, say, the time at the beginning and end of the period of interest. This, as well as other purposes, could be served by computing a *definite* integral of the model. Continuing with the example above it might be of the form $\int_a^b x^n dx$, where the integral is taken over the interval between the limit points a and b

on the x-axis. Performing the integration one way or another, the desired process value becomes available from the difference between the values of the definite integral at both terminal points, namely, $F(b) - F(a)$, where F is common notation for a function f that has been integrated. Popular in modeling real-world processes are definite integrals over time between the beginning, t_0, or simply 0, corresponding to point a, and a time later on, t, corresponding to point b and given as $\int_0^t x^n dx$ which equals $F(t) - F(0)$.

Note also that: (1) derivatives and integrals of certain functions are used often and are available in tables; (2) between the two operations, obtaining integrals is, in general, less straightforward than derivatization, but applicable to a wider range of functions; and (3) with many and important real-world models, neither operation may be possible or practical. There are, however, alternatives, as will be seen next and in Chapter 3, Alternative infrastructure/series, numerical methods.

Coefficients and time

Taken to represent an essential relationship between a process and its governing factors, mathematical functions obviously have a central role in processes modeling. The actual course of a process, however, will also depend on the *coefficients* that specify absolute values (see above). Their role is not trivial. In fact, the makeup of the coefficients will have a major role in determining collective behavior patterns in more complex processes later on.

Modeling the real world, cellular processes included, often deals with and takes for granted that rules such as rate laws stay constant during a process. Model equations in these cases employ so-called *constant* coefficients (i.e., all coefficients are constants). In the literature and here, they are often assigned the symbol a, but other symbols are used as well (see Eqs. 1.1a,b, 1.2, and 1.5). A constant coefficient was, in fact, already encountered above using the symbol C for the (constant) number of ATP molecules available from glucose. Consider metabolite synthesis in cells that at the same time accumulate deactivating mutations. To accommodate a secondary process such as this, affecting only the amount, the numeric value, say, of metabolite produced, one can employ *variable (nonconstant, unrestricted)* coefficients denoted, say, $a(t)$. These depend directly on the time—or another independent variable—and generally burden the mathematics.

Operators as facilitators and delimiters

A concept that can have meaning of its own and is practical in simplifying the handling of *DE*s is that of an *operator*. More advanced, it comes with an economical notation. Seldom in this book, but frequent in the literature, operators here will be symbols for an instruction to perform a particular mathematical operation on a function. Such might include "multiply by one (or any other number)," "raise to the nth power," "integrate," and so forth. Especially relevant here would be the *differential operator*, D (also d/dx, or d, among others) where Dy means "take the derivative of the function y (with respect to x)" so that $Dy = y'$. Similarly, $D^2y = y'' \ldots D^{(n)}y = y^{(n)}$, with superscripts denoting the order, the number of times a derivative is to be taken. More complex operators are also in extensive use with a few examples in this chapter and elsewhere in this book. Operators have associated theory on how and where they can be applied. Among others it may help find representative functions for a problem and define limits for their application.

Choosing a differential equation model

Options

Modeling cellular processes largely focuses on change at the molecular and near higher levels. Its major theme currently is how processes respond to various driving factors, mainly over time. Often referred to as process dynamics, the literature generally uses for this type of modeling the essentially synonymous term *kinetic* modeling. Taking advantage of what *DE*s, linear and nonlinear, can offer, *DE* models are considered the gold standard of process modeling and predominantly belong to one of three equation prototypes. These offer an increasing level of complexity of what can be represented and will be introduced next, beginning with the simpler prototype.

Single variable dependent only/ordinary differential equations

The synthesis of a cellular protein is often portrayed by a barebones single term *DE*, such as $dP(roduct)/dt(ime) = v(elocity)$, where P is the amount/concentration of protein, t is the time, and v is a constant representing an unchanging rate law. Here, P is the dependent variable (generically y) and t the independent variable (standing in for x), as shown by Eq. (1.1a). In the cell, the protein will subsequently be degraded, as represented again by a single term equation, except now using a different formula, $dP/dt = -\alpha P$ where α is another rate-related constant specific to this reaction, as shown in Eq. (1.1b).

$$\frac{dP}{dt} = v \tag{1.1a}$$

$$\frac{dP}{dt} = -\alpha P \tag{1.1b}$$

Biochemical (and other) change in living cells or in the test tube is often described as a balance between the concentrations of materials entering a reaction and those disappearing. A more complex process with multiple *partial-* or *subprocesses,* it can be represented by equations with multiple corresponding terms. To describe total protein levels in a system, the simultaneous synthesis and degradation of induced protein may thus be combined into one equation.

$$\frac{dP}{dt} = v - \alpha P \qquad (1.2)$$

An equation such as Eqs. (1.1a,b) and (1.2), which describes a process as it depends on a *single* independent variable, the time or other, is called an *ordinary differential* equation (ODE, plural *ODEs*). Widely applied in modeling real-world systems and their processes, *ODEs* are also a mainstay of the infrastructure for dealing with cellular behavior.

Building blocks such as those in Eqs. (1.1a,b) and (1.2) are quite common in cellular process modeling with *ODEs*. However, the choice is, offhand, much wider. Major options for *linear ODEs*, often preferred in modeling, can be seen in Eq. (1.3), now written in a somewhat different way (with more on that below).

$$a_0(x)y + a_1(x)y' + a_2(x)y'' + \cdots + a_{n-1}(x)y^{(n-1)} + a_n(x)y^{(n)} = Q(x); \quad y(x_0)$$
$$(1.3)$$

With the symbol y being given by any appropriate function, building blocks may include also higher derivatives, as indicated by the superscripts. Each term may have a coefficient of its own. These may be variable, $a_i(x)$, as shown here, or constants, a_i. As used often, the subscript i in a_i is an *index* that refers to any one particular member in a list, here of terms. Lists such as these have n members, each with its own index adding when convenient also a 0 term.

Two *ODE* defining features are especially noteworthy. First, the *order*, n, is the highest derivative that appears in the equation. Equations of order one (i.e., with the term dy/dx or y') but no higher derivatives, are *first order* and are usually more manageable than those of higher order. The latter, however, may first be converted into more readily solvable systems (see below) of n first-order equations. As it were, *DEs* representing cellular processes are rarely higher than second order. Second, the *degree* of a DE is the greatest *power* of a term with the highest derivative. Higher power terms, absent in Eq. (1.3), would have required special notation. They were, however, deliberately omitted because that would have made the equation nonlinear, nonadditive, and more problematic. In these terms, Eq. (1.3) is thus of the nth order and of degree one.

Having dealt so far with the LHS of Eq. (1.3), the $Q(x)$ term on the RHS is a mixed blessing. In modeling the real world it has a major role, notably in representing extra inputs into processes that are already on their way. This includes, for example, additional precursors in a material process and typically the enhancing or negative effects on process activity by control. Mathematically, the $Q(x)$ term is a spoiler of sorts. It has the distinction of being allowed to consist of one or several functions *other* than y, but that still depend on the same independent variable x (or t), as well as simply be a constant, Q. Absent $Q(x)$ or Q, the observed variable y is present in all terms and the equation is *homogeneous*—often represented by putting the RHS of equation such as Eq. (1.3) to equal zero ($=0$). By the same token Eq. (1.3) is *non/inhomogeneous* when the similarly named nonhomogeneous term(s) $Q(x)$ or Q are present. It turns out that dealing with homogeneous *ODE*s, can be relatively straightforward. On the other hand, increasing the scope with $Q(x)$ terms can make subsequent handling significantly more difficult, as will become evident later in this chapter and in Chapter 4, Input into output/systems, where it is common in technological process models. The difference Q terms make is not trivial. As readers well know, the steady state of cellular processes in which throughput is constant and there is no net change in the system is a major phase in routine cellular existence—a good biological reason to model the steady state. There is also strong mathematical incentive. Process rate equations often can be put in the form of Eq. (1.3), keeping the coefficients constant and omitting all terms on the RHS except the first two. When there is net change in the system, its rate is represented by a Q term with its consequences. At steady-state throughput stays constant, the rate term equals 0, and Q disappears to make for a favorable homogeneous equation. As such the steady state is also a unique and well-defined condition, not the case retaining the Q term which, offhand, can take any physically reasonable value. Altogether, modeling at steady state, as will be seen, provides the mathematics with access to process rate parameters that are useful not only at steady state, but throughout processes' entire history.

In terms of the real world, equations with constant coefficients and a constant Q term, but not an x-dependent nonhomogeneous term, if any, correspond to systems whose rate of change laws stay constant during an observation. Both system and equation are then known as *autonomous*, a highly desirable property common in modeling and predominant in this chapter.

As for equation *format*, there is considerable freedom where and in what order to place terms in *ODE*s or other *DE*s, as long as the mathematics is preserved. Obviously, Eq. (1.2) differs in appearance from Eq. (1.3). However, it should not be difficult to see that Eq. (1.2) consists of counterparts to the first two terms and the Q term of Eq. (1.3) where: (1) the symbols y and x were replaced with P and t, and y' by dP/dt; (2) the coefficients were constants that equal α and 1, and Q became v; and (3) rearranging the terms by subtracting from both sides—αP will yield $\alpha P + dP/dt = v$—Eq. (1.2) in the format of Eq. (1.3). Note also that given a model with subprocesses such as Eq. (1.2), some contributions may become irrelevant in particular conditions, say, when protein degradation is very slow and the αP term is, therefore, very small. As already pointed out, it could be safely omitted, but the equation remains valid and still subject to the usual rules.

Taken together, *ODEs* evidently offer a versatile modeling tool—be it in the choice of functions, their order, and degree, as well as because of coefficients that can be with or without restrictions, and the various options available for nonhomogeneous terms. In practice, as often elsewhere in real life, the more descriptors required for a process, the less effective a description is likely to be. Similarly, the more the *ODE* options used, the less the mathematics is likely to perform. Higher order derivatives may not exist. Integrals of many functions remain unknown. Seemingly minor differences may be detrimental to further manipulation, for example, when constant coefficients become variable or when equations include nonhomogeneous terms. Frequently an only workable choice, it is not unusual to adopt or adjust real-world conditions so as to allow modeling by the simplest of *ODEs—first order, linear, homogeneous*, and with *constant* coefficients.

Given a realistic choice, the makeup of *ODE* models at this point could meet the above first objective, finding a general rate law. To obtain process values unique to a particular setting, the second objective, the mathematics requires not only the right equation but at least one particular process value, a so-called *initial value* or *initial condition*. Some problems are defined using both initial and final, so-called *boundary values* or *boundary conditions*. However, also known as *initial value problems*, *ODE* models in this chapter will commonly rely on initial values or conditions, as given by the lone term on the far RHS of Eq. (1.3), $y(x_0)$. It is the value of y, call it y_0, at the initial point of reference namely x_0. It could be time zero of an experiment or the initial concentrations in a biochemical process, for some readers the familiar baseline. Even if equal to zero, initial values have to be provided because of the mathematics. Obviously, they are also a link between the model and the real world. Initial value models, incidentally, are required to meet not only existence and uniqueness conditions, but also to vary continuously with change in the initial conditions.

And finally, note in a more general vein that in addition to actual applications in modeling, *ODEs* provide direct tools as well as general ideas on representing and investigating change for other types of real-world models—whether in modeling with partial differential equations (PDEs), as will be discussed next, or turning to difference equation models later on. As it were, *ODEs* also have an increasing role in models presenting organizational features as will be discussed in Chapter 8, Organized behavior/network descriptors.

Multiple variable dependent/partial differential equations

In modeling with *ODEs* so far, it was tacitly assumed that processes took place at the same rate everywhere in the system. In the real world, however, location can matter. In cells, substance mobility may be sluggish—free diffusion is slow and there can be membranous barriers, among other retarding factors. Precursors could consequently be in short supply; inhibitory products could accumulate. It might affect the activities of particular cellular components as well as the overall throughput, the flux, of larger cellular pathways, metabolic, signaling, or others. A ubiquitous model for intracellular flux is the *reaction diffusion* model. Wide-scoped substituting factors, the same model also can describe the propagation of nerve impulses, chemotactic mobility in microorganisms, morphogenesis in embryos, or the migration of humans, among others.

When applied to the flux of a cellular component, the reaction diffusion model employs the factors listed in the provisional model above to describe: (1) the amounts and where in the cell the component is located as given by the concentration, X, the space coordinates, x, y, z, and a diffusion constant, D, an index of the component's mobility; and (2) the change in total substance concentration depending on the various biochemical reactions, here represented by their rates, v. In actual detail, as seen below, the model calls for multiplying the diffusion constant of substance x, D_x, by the change in the *gradient* of the concentration, X, as it varies along each of the dimensions in space, x, y, and z. The concentration gradient, the driving force, is given by the *derivatives* of the concentration with respect to the coordinates. As for the reactions, ongoing synthesis and breakdown of cellular substances will be represented by the total concentration as it relates to the time (i.e., by the sum of the appropriate reaction rate parameters, v_i, the subscript i again serving as a label for each particular biochemical reaction).

Represented by an equation, flux depending on the time and coordinates is a single dependent variable that is a function of several independent variables. Calling for partial derivatives and retaining the symbols from above, the reaction diffusion model is in the form of a *PDE*.

$$\frac{\partial X}{\partial t} = D_x \left(\frac{\partial^2 X}{\partial x^2} + \frac{\partial^2 X}{\partial y^2} + \frac{\partial^2 X}{\partial z^2} \right) + \Sigma v_i \tag{1.4}$$

Eq. (1.4) is second order, that is, it uses *second* partial derivatives. That is so because what changes in the process and will be derivatized is not the concentration as usual, but preexisting local concentration gradients that already are represented by a derivative, say, $\partial X/\partial x$. As for the symbol Σ, the Greek letter *sigma*, it commonly denotes *summation* in discrete (discontinuous) representations, discussed later. It is legitimate in this otherwise continuous model since the biochemical processes represented are indeed continuous, it is the particular rate parameters that are usually discrete and offhand do not affect the nature of the process.

*PDE*s currently have a respectable but limited role in modeling ongoing cellular processes systems biology is concerned with, as they will have in this book. *PDE*s will have a significant role in evaluating process optimality, as will be seen in Chapter 6, Best choices/optimization. *PDE* mathematics, however, is generally advanced and detailed, and here largely out-of-scope. A short survey of basic properties, classification, and handling should fit the bill.

With building blocks and operations in common with *ODE*s, there is natural overlap in mathematical properties and equation handling, as well as carry over into terminology and classification. Turning to notation first, *PDE* literature keeps some of the *ODE* conventions, but also has some of its own. The letter u thus often serves as a generic symbol tor the dependent variable, an observed property such as the flux. Also common is replacing standard partial derivative notation, for example, $(\partial u/\partial x)$, $(\partial^2 u/\partial x^2)$, or $(\partial^2 u/\partial x \partial y)$, where u is the dependent and x and y the independent variables with a more convenient and concise notation, for example, u_x, u_{xx}, and u_{xy}, respectively.

What types of *PDE*s are likely in cellular modeling? Common classification is based on the order, the highest derivative(s) present. As it were *PDE*s of the first and the second order are the most common and mathematically manageable in modeling the real world. As with *ODE*s, models are preferably linear; nonlinearity in nature and in a model is problematic here as well.

First-order partial differential equations

In their most general version, with one dependent variable, u, and two independent variables, x and y, first-order *PDE*s are in *quasi-linear* form, shown here in subscript notation.

$$a(x, y, u)u_x + b(x, y, u)u_y = c(x, y, u) \tag{1.5}$$

Here a, b, and c, are coefficients that, as with *ODE*s, may be constants or dependent on any one or combination of the variables (in parenthesis). Further classification of first-order *PDE*s usually depends on whether and which terms harbor the dependent variable u. Similar to *ODE*s, when u is present in all terms the equation is homogeneous and nonhomogeneous otherwise.

A first-order quasi-linear *PDE* applied in modeling biological processes is the *advection* equation. It can apply in modeling, among others, substance uptake by cells from a flowing medium. When substances at the same time also react and change in concentration, the advection equation, not shown here, is similar in format to Eq. (1.4). However, the movement of the substances depends on bulk flow velocity represented by a constant of its own instead of diffusion constant and the dependence on location is given by the first-order partial derivatives $\partial X/\partial x$, $\partial X/\partial y$, and $\partial X/\partial z$, rather than the second-order expressions above.

Second-order partial differential equations

Many a natural and technological process can be represented by second-order *PDE*s extending beyond the Eq. (1.4) prototype. Again, in subscript notation and reminiscent of Eq. (1.3), the linear version in its most general form is notated, where the coefficients a through g are constants or functions of x and y.

$$au_{xx} + bu_{xy} + cu_{yy} + du_x + eu_y + fu = g \qquad (1.6)$$

Now the quantitative properties of processes represented by both first and second-order linear *PDE*s as a rule are dominated by the highest order terms. Taken together these are the *principal part* of the equation. For example, the first three terms on the *LHS* of Eq. (1.6). Second-order *PDE*s are commonly classified based on the coefficients of the principal part. More specifically on the expression $b^2 - ac$, the *discriminant*, D, and whether it is less, or more, or equals zero. Named after trigonometric curves whose formula discriminants obey the same relationships, the main prototypes in describing nature are shown here.

$$\frac{\partial^2 u}{\partial x^2} + \frac{\partial^2 u}{\partial y^2} = \quad D < 0 \quad Elliptic \qquad (1.7a)$$

$$\frac{\partial^2 u}{\partial x^2} - \frac{\partial^2 u}{\partial y^2} = 0 \quad D > 0 \quad Hyperbolic \qquad (1.7b)$$

$$\frac{\partial^2 u}{\partial x^2} - \frac{\partial u}{\partial y} = 0 \quad D = 0 \quad Parabolic \qquad (1.7c)$$

As can be seen, the prototypes are related and occasionally are interconverted on purpose. Chameleon-like, if you will, certain *PDE*s, in fact, change their colors and switch behaviors on their own, depending on the environment or the particular interval.

What prototype does Eq. (1.4), the ubiquitous reaction diffusion model, belong to? Readers who opt for Eq. (1.4) being parabolic are correct. It is directly related to Eq. (1.7c), which is also known as *heat* or *diffusion equation*. The two other prototypes above have major applications, as well. Obviously related to the diffusion term, first on the *LHS* of Eq. (1.4), but here *elliptic*, is the *Laplace* equation, $\partial^2 u/\partial x^2 + \partial^2 u/\partial y^2 + \partial^2 u/\partial z^2 = 0$, where x, y, and z, are again the coordinates in space. As such it may represent the flow (e.g., diffusion) within a system of a conserved quantity, be it a cellular component, a fluid, or heat. Elliptic *PDE*s serve in modeling the intracellular and, in some cases, the intercellular movement of particles or signaling molecules. The remaining *hyperbolic*, so-called *wave* equations $\partial^2 u/\partial x^2 - \partial^2 u/\partial \tau^2 = 0$, indeed describe wave motion, and could, for example, be models for the propagation of electric signals, such as the action potential in axons. That wave equations are at the center of the quantum view of matter is one of their major glories.

Prominent *PDE* models employ boundary as well as initial values. Boundary values can be simple constants or changing values as given by an equation. As with other *DE*s, these ancillary relationships link the representation to the real world. With the classical case of the fixed end points of a vibrating violin string, here relevant would be, for example, cellular boundaries between which substances will diffuse in reaction diffusion models, or time limits in which a technological or biological control is allowed to accomplish its task (see Chapter 6: Best choices/optimization). As it were, *PDE* boundary values and how they are represented can have a major role in subsequent workup and what route it will take.

Altogether *PDE*s are favored to be *well posed*. That means that they are expected to have solutions and such that are unique. The solutions of a model should also respond continuously and reasonably proportionally to minor deviations in the initial or boundary conditions, say, natural variations in setting up an experiment. Observed also with *ODE*s, but more common with *PDE*s, being *ill* posed is problematic, conducive to computer error, and may call for revising or replacing an original model.

Unity in diversity/systems of ordinary differential equations

Major functions in nature, as well as in the artificial world, are realized due to complex processes in heterogeneous systems. When diverse system members act collectively it is likely that there will be some accommodation between members in which not all needs will be satisfied and not all potentials realized. Collective behavior, the whole, is then a *consensus* of the behaviors of the parts, the contributing processes and their interactions. Each process has its share of the total activity, while also satisfying the others. What in a complex process will be a consensus? Can it change? How? By what? Cellular functions combining various lower level processes arguably can be seen in similar terms. In metabolism, gene expression exchanges with the surroundings, to mention a few.

A complex whole/no parts

Much of what is already known about collective behavior in cellular systems at various levels is based on simple observations. All one asks of a process, no matter its workings, is how what comes out relates to what went in. It has been successful with elementary chemical reactions such as that of glucose and hemoglobin and with increasingly complex processes such as enzymes converting substrate into product (aggregate protein metabolism represented by Eqs. 1.1a,b and 1.2) up to high-level cellular processes such as viral replication, among many others. How output relates to input is usually also the main concern in various applications, notably bioassays that measure enzyme activity, chemical mutagenesis, or microbial infectivity. Modeling these processes with *ODEs*, they are represented by single, lone-standing, so-called *scalar* equations. These can vary with process properties delineated above, typically depending on whether or not the process is linear, maintains a constant rate throughout, and receives input while in progress.

The complex whole/with parts

It is this type of model that addresses the major question in systems biology raised in the beginning of this chapter. How do the quantitative behaviors of complex processes, the whole, relate to those of the lower level components, the parts? And vice versa, how do the parts, the subprocesses collectively make for the whole? Given that cellular systems and their partial processes are diverse, interact and share resources, their collective behavior can be taken to be a consensus such as described above. The immediate modeling objective is to establish values that, at the same time, satisfy both the overall process and the partial processes, collectively.

To proceed, however, it will be useful to look first into how collective behavior in complex cellular processes might be seen for modeling purposes. In metabolic or signaling pathways, for example, early members supply intermediates to those downstream. Latter steps, in turn, may affect earlier ones. They could reverse or alternatively stimulate the reaction of the immediately previous step directly. And they might provide longer range, positive or negative feedback to steps higher upstream. In more comprehensive models, a pathway also interacts with other cellular or external systems such as those that trigger activity, supply or accept building blocks, or serve as energy sources or sinks. Still higher level cellular interactions, say, in gene expression, whole-cell differentiation or response to external stimuli often can be seen in similar terms. So would interventions, chemical and genetic, in medicine, agriculture, or industry.

With pictures such as this in mind, the currently predominant modeling strategy is *bottom-up*—reductionist, if you will. In line with much else in science, one tries to use the parts to describe the whole, the partial processes to construct the more complex biological function. In a less common and complementary *top-down* strategy (described in Chapter 5, Managing processes/control), the behavior of the combined process is interrogated to establish roles for the subprocesses. Obtaining data and creating realistic descriptions of complex processes is a major concern in cellular process modeling of all kinds. There are already models of complex processes, albeit few, that can provide insights into how cellular control may work or predict how cellular processes would respond to, say, mutations.

In the bottom-up strategy, each subprocess would be represented by an equation of its own with its particular dependent and independent variables within a *system of (simultaneous) equations*. Favored in cellular modeling are systems of, once again, *ODEs*. For example, in the form of equation system (1.8), models can represent *ongoing* processes as well as more *permanent* process features. What follows will emphasize infrastructure for modeling ongoing processes; other purposes will be covered later on in this chapter and elsewhere in this book.

Modeling cellular processes with systems of *ODEs*, two bottom-up options take center stage. One, common in creating cellular process models, is eclectic in its building blocks and methods and will be referred to as *customized* modeling. The second, well-established in technological applications and finding its way into cellular modeling, is based on specialized linear mathematics and here will be called *structured* modeling.

Customized modeling

Cellular process diversity is naturally reflected in their different quantitative and time behaviors. At this time there is no general model for all known behaviors, let alone those that still may be discovered. Turning instead to individual cases, customized modeling is a long-established and major force in exploring ongoing cellular processes. Customized models initially aim for maximum fidelity and a wide-scoped portrayal that can capture the greatest essential detail. Models of interest aim primarily to describe process trajectories and throughput of entire pathways as well as those of the reactions of single or teams of pathway members. Such models might provide process values and rates for immediate practical or theoretical application and, hopefully, long range insights into biological function that depends on the process. Often the immediate interest is in possible effects of change in system makeup, say, due to mutations, or in its response to change in its environment, perhaps in substrate concentrations.

An inclusive approach, customized models can employ a mix of *ODEs*, allow linear and nonlinear versions and possibly admit also other equations, here mostly algebraic. Modelers can manipulate and combine particular equations individually. As they appear in the literature, traditional customized models may already have been reduced in size by initially including only process steps thought essential to the dynamics and possibly simplified mathematically, as well.

The relative freedom to choose and handle expressions often results in demanding mathematics, nonlinear equations to begin with. Cellular processes often include components that interact directly: nucleic acid sequences and proteins in gene expression, enzymes that bind substrate, *ES*, or various effectors and cooperative binding of oxygen to hemoglobin are among many others. Sometimes nonlinear models can be handled by linearization or other means (see linearity and nonlinearity, above) early on, possibly before a model is in its published version. As will be seen later on, expressions that are difficult to handle or compute are often a major issue, among others.

When modeling large pathways, additional factors may come into play. A significant component in these models can be *constraints*. These can represent limitations on processes imposed by nature, notably on how far and fast reactions can go, as well as the need to conserve matter. Constraints in that sense are a link to the real world that can narrow the scope of subsequent handling of the model. Constraints can also satisfy purely mathematical needs, as discussed later on. They can be represented separately, possibly by algebraic equations.

In a format extending on biochemical kinetics, examples of customized models are readily available in the textbooks listed in the Introduction. An *ODE* system model of small cellular control devices, albeit in conjunction with other modeling tools, is discussed at some length in Chapter 8, Organized behavior/network descriptors (see Eqs. 1.2−1.5 and accompanying text). A conventional, "pure" *ODE*, only counterpart is available in Ref. [1].

To gain some insight into what considerations might go into creating and choosing infrastructure for large customized models, early literature can be helpful. As it were, the glycolytic pathway is a longtime favorite in kinetic modeling, a standard testing ground for new methods. The Rapoport et al. model of the flux (throughput, turnover) and other process properties of that pathway [2] is here out-of-scope in its full detail and advanced methods. But a few points about the model are notable: (1) Glycolysis is the pathway that converts glucose into pyruvate; the latter can be converted further into lactate. (2) Human erythrocytes provide in vivo conditions favorable for a quantitative model of the process. It is the sole energy-producing pathway, it is at steady state, enzyme concentrations stay constant. There are also no significant barriers to the movement of metabolites, which allows for a common assumption that the system is *well stirred/mixed*. That is, substances can react freely, irrespective of physical limitations. (3) There is a hierarchy of rate behaviors. Some of the pathway steps are reversible and fast and can be considered as equilibria between metabolites that essentially stay constant. Others are irreversible, relatively slow, and presumably dominate the overall flux. (4) The model represents a steady state of 15 enzymes that participate in the glycolytic pathway. It includes also the conversion of pyruvate into lactate, as well as associated reactions of adenine nucleotides and NAD^+/NADH. Glucose, the only precursor, is taken to be supplied continuously and in an amount that would allow each enzyme to be maximally active. (5) Conventionally, each enzymic step would be represented initially by an *ODE* rate expression of its own. Based on rate properties of the pathway's reactions, the model can be reduced to its predominant (irreversible, slow) key steps. It can be reduced further by combining certain pathway steps together, and eventually ends up with four initial rate equations. Modeling pathway behavior takes into account these equations, equilibria between intermediates of the fast reactions, and the need to conserve total amounts of adenine nucleotides and $NAD^+/NADH$. (6) Change in metabolite concentrations with time, the dependent and the independent variables in these equations, is described in terms of pathway member-specific rate expressions. These include: (1) nonlinear, second-order rate constants, familiar from chemistry; (2) linear first-order with and without zero-order rate constants; as well as (3) expressions that could be linear or nonlinear such as the Michaelis—Menten formula and a first-order rate constant expression that includes inhibition. (7) A major goal of this model was to identify pathway steps that may control the overall (steady state) flux. As such, the model is an early example of what is now known as *metabolic control analysis*. Of no concern here, a more general discussion of this approach to modeling cellular processes is available in Chapter 5, Managing processes/control.

Often larger, more recent and contemporary models with similar objectives, such as the more complex cell cycle control process in budding yeast and elsewhere [3], take into consideration the timing of certain events. These models are computer implemented from the very beginning, employing a variety of largely standardized software packages where modelers input relevant parameters such as substance concentrations or rate expressions. These routines choose and specify the mathematics that will be used to describe and compute collective behavior of process participants with more detail later on. Although contemporary procedures can accommodate on the order of one hundred subprocesses, model size remains a significant computational issue. Arguments such as those empirical ones, used to reduce the size and simplify the glycolysis model, can now be applied quantitatively in standardized ways using *model reduction* routines.

Taken together, there is substantial success in customized modeling of large pathways, but it has been limited so far to a few data rich systems. As it were, needed experimental data are generally in short supply. Applying data to an in vivo model of, for example, signal binding to a receptor, enzyme activity, or transmembrane transport that have been obtained in isolation and in the test tube may be questionable. The strategy is subject also to other limitations—rate laws are not sufficiently known, equations can be difficult, creating, managing, and computing large models becomes increasingly demanding. Customized models of complex processes are typically open to only limited formal mathematics and usually end up being evaluated not as a formal solution but by alternative, mostly numerical methods (see below and Chapter 3: Alternative infrastructure/series, numerical methods).

Structured modeling

Major processes in nature and in the artificial world exhibit *linear* behavior on their own or can be represented as such using mathematics. This creates opportunities for turning to a different, strictly linear modeling approach in which processes will be represented by systems of linear *ODEs*. This approach relates lower level contributions to complex processes in a more systematic, standardized, and, in effect, *structured* format. While the customized approach allows for some freedom in processing its models, including handling single or small groups of equations separately and differently, the structured format workup will be collective, handling all equations simultaneously and similarly. Compared to customized modeling, this approach is more restricted in its building blocks and narrower in the scope of processes that can be described. But there are also substantial benefits. Models have formal solutions, allow rigorous formal analysis, and offer unique information. Functional relationships between systems and processes (see Chapter 2: Consensus/linear algebra and Chapter 8: Organized behavior/network descriptors) can be represented explicitly and kept track of. Altogether, structured models are easier to manage, scale up, and are computer-friendly.

Taking up structured modeling, less likely to be familiar to readers, will serve several purposes here. Its mathematics has been, and continues to be, instrumental in modeling cellular processes, being also a point of departure for characterizing more permanent, higher level organizational features of cellular network processes especially significant in biotechnology. The structured approach is also a mainstay in various sciences and is becoming central to modeling technological system processes and control with biological counterparts to be taken up in Chapter 4, Input into output/systems, and Chapter 5, Managing processes/control. Here, structured modeling will be an opening to the methods of *linear algebra*—the underlying mathematical discipline that is devoted to dealing with equation systems and taken up in some detail in Chapter 2, Consensus/linear algebra. Valuable in later-stage computing of customized models, it has wide applications in a variety of modeling elsewhere. And last, but by no means least among the purposes here, is that the structured approach offers an opportunity to show that what can matter in cellular and other complex processes is not only their nature, but also their quantitative makeup and its power to determine process behavior patterns.

Creating a model, once again, begins with representing subprocesses by separate equations, here first degree *ODEs* exclusively. Their building blocks will be terms of Eq. (1.3) where the first-order derivative, second on the RHS, is placed on the LHS as shown below. The RHS will have the first term of Eq. (1.3) remain on the RHS to represent the dynamics of the particular subprocess. Depending on the case, there could be more terms such as this on the RHS if there are contributions from and into other processes. Nonhomogenous Q terms may be included as well. In this format, seen in Eqs. (1.1a,b) and (1.2), the *ODE*s are assembled into systems of simultaneous equations. To obtain a formal solution for a system with n variables will take representing each by an equation of its own making for a system total of n simultaneous equations. To reach that number, incidentally, is the frequent mathematical reason for including constraints in a model mentioned above and discussed more below and in Chapter 2, Consensus/ linear algebra.

A common and general linear *ODE* system prototype with n dependent variables, $y_1 \ldots y_n$, and one independent variable, x (or t), takes the form:

$$
\begin{aligned}
\frac{dy_1}{dx} &= a_{11}(x)y_1 + a_{12}(x)y_2 + \cdots + a_{1n}y_n + Q_1(x) \quad y_1(x_0) \\
\frac{dy_2}{dx} &= a_{21}(x)y_1 + a_{22}(x)y_2 + \cdots + a_{2n}y_n + Q_2(x) \quad y_2(x_0) \\
&\vdots \qquad\qquad\qquad\qquad\qquad\qquad\qquad\qquad\quad \vdots \\
\frac{dy_n}{dx} &= a_{n1}(x)y_1 + a_{n2}(x)y_2 + \cdots + a_{nn}y_n + Q_n(x) \quad y_n(x_0)
\end{aligned}
\tag{1.8}
$$

where the coefficients are indexed by letters i and j, which provides each term with an address denoting row (horizontal) and column (vertical) locations within the system.

As with managing their single equation members, the more practical *ODE* systems are autonomous (i.e., with constant coefficients, a_{ii} and a_{ij}, and with no Q terms). Obtaining particular solutions requires the initial values—here entered in the separate column on the extreme RHS. Solving *ODE* systems such as this using linear algebra, one initially groups the coefficients and the variables into separate structures, matrices and vectors. In this format equation system (1.8) is represented by Eq. (1.18a), much more compact and convenient and often used initially instead.

Many complex processes combine independent contributions. These can be represented in the Eq. (1.8) format with each row standing for that particular subprocess only using the first term on the RHS. When there is no input from other subprocesses and no additional RHS terms the system is said to be *decoupled*. In the presence of other contributions, represented by terms with unequal coefficient indices, that is, $i \neq j$—the system is *coupled*. The latter here means that one needs y_1 to know y_2 and all other terms, and so forth. It turns out that the mathematical handling of such equation systems will differ. Decoupled systems, simpler, are also the easier to manage.

Existence and uniqueness

Setting up *ODE* equation systems such as system (1.8) and others, one expects that there is a solution and aims for one that is unique. In other words, that terms in each equation relate according to the rules delineated earlier and that together they indeed can generate one, and only one, consensus value that will satisfy all the variables simultaneously. In assembling a useful equation system, it matters also how equations relate.

Two points are noteworthy. First, as a general rule, equations have to be *consistent* in order for solution values to represent a consensus (i.e., to satisfy multiple equations at the same time). If an RNA polymerase molecule is to form a transcription complex, its path inside a cell has to share points with those of associated transcription factors (i.e., the equations that describe their locations have a value in common). This would not be the case if they all followed parallel paths whose equations are not consistent. Second, given consistency, obtaining a meaningful and unique solution hinges on equations, not only the terms, being *linearly independent* of the others. It turns out that this is a major issue in modeling and will be taken up in some detail in Chapter 2, Consensus/linear algebra. Recall here that terms of a single equation are linearly independent when they are dissimilar to the point that they cannot be interconverted by changing numerical constants. When it comes to entire equations in a system, such as the rows of system (1.8), they will be linearly independent when one row cannot be expressed in terms of one or a sum of the other rows, whether or not they were multiplied before by a constant. Problems can arise when, for practical purposes such as improving accuracy or representing constraints, real-life equation systems are set up with more than n equations for n variables. Or else when, due to data limitations, they are fewer. Being *overdetermined*, or *underdetermined*, respectively, usually call for special methods left to more specialized sources.

Note also that: (1) Systems of simultaneous *ODE*s arise also for purely mathematical reasons, for example, as a first step in solving higher order, single *ODE*s as well as various single *PDE*s. (2) There are also real-world models as well as mathematical methods that employ systems of *PDE*s. (3) Some systems of nonlinear *ODE*s display the dramatic behaviors of *chaotic* systems, present also in cellular systems, these will be left to more specialized sources elsewhere.

As a prominent example of modeling cellular collective behavior with an *ODE* system, there is so-called *stoichiometric* modeling. In a format comparable to *ODE* system (1.8) and open to linear methods, it could be used, for example, for the flux through the four initial steps in glycolysis in a linear model (see Ref. [4]). Specifically the reactions glucose$_1 \rightarrow$ glucose-6-phosphate$_2 \leftrightarrow$ fructose-6-phosphate$_3 \rightarrow$ fructose-1,6,-biphosphate$_4$ where $\rightarrow$ and $\leftrightarrow$ stand respectively for irreversible and reversible steps and the subscripts $i = 1, \ldots, 4$ index the pathway members. The dynamics of the process, the change over time in its member concentrations, can be expressed by *ODE* systems (1.9a−d).

$$\frac{dS_1}{dt} = n_1 v_1 \qquad (1.9a)$$

$$\frac{dS_2}{dt} = n_2 v_2 \tag{1.9b}$$

$$\frac{dS_3}{dt} = -n_{-2}v_{-2} \quad n_3 v_3 \tag{1.9c}$$

$$\frac{dS_4}{dt} = n_4 v_4 \tag{1.9d}$$

As can be seen, the reaction rate of each pathway member, i, is given by the derivative of its concentration, S_i, with respect to time, t, entered in consecutive rows on the LHS. The rate of each step depends on the number of substrate molecules that participate, the stoichiometry of the reaction, n_i, and on the rate, v_i, as given on the RHS. Each step here is carried out by the forward reaction of the respective member as represented by a positive term on the RHS. The reversible third step features also the back reaction, represented by a separate, negative term. All stoichiometric coefficients, n_i, here equal one (see also Eq. 2.5 in Chapter 2, Consensus/linear algebra). Since there are no inputs or outputs by other processes, other terms are omitted. In pathways elsewhere, substances later in the pathway may enhance or inhibit a reaction a few steps before and would be represented by a term with an appropriate sign, say with a minus if the feedback is negative (see also Eq. 1.8 and accompanying text).

With all their differences there is also a noteworthy common theme in how the customized and structured modeling approaches are being applied in cellular modeling. Both approaches are most often used to model cellular process at steady state, common also in modeling other real-world processes; it was the case with the glycolysis flux example above and it will be the case repeatedly with structured models in this chapter and elsewhere in this book. With biological and linear mathematical incentives noted previously, the steady state can also serve as a condition that allows to turn nonlinear *ODE* models linear and open to the structured approach. It was used this way for example in early modeling of gene expression based on the synthesis and degradation of m-RNA and proteins. Processes at steady state are also the point of departure for more specialized applications such as establishing preferred pathways in metabolic networks among others.

And, finally, there is still another, more general consideration that can become significant in setting up cellular process models. The mathematics one chooses should, obviously, best represent the process and the purpose(s) of the model. However, in contemporary modeling, certainly of complex cellular processes, the mathematics is generally no longer on its own. Modeling procedures routinely turn to computer computations. In that mode, certain mathematical expressions and operations can be considerably more practical than others. Practicality can be a major factor in choosing what and how mathematics will actually be used with a model (more in Chapter 3: Alternative infrastructure/series, numerical methods).

Description into general law, interpreted

Solving equations

Given *DE* models that were created according to the rules, how will the desired information become available? For one, it will hinge on two notions that underlay modeling processes with *DE*s. The first notion is that change observed in a particular setting is an expression of a more general law, possibly still unknown, that reflects a property of the process and can determines process behavior in other conditions as well. Obtaining the general law can be instructive in its own right in studying underlying mechanisms. Often the primary goal in obtaining the general law is to provide, after further processing, numerical process values. These frequently meet practical needs in ordinary life, here they will characterize a process and can also serve theoretical purposes, say, in the analysis of how process characteristics depend on the quantitative parameters of a particular process, as will be seen later. The second notion underlying modeling with *DE*s is that equations of an original model representing actual change hold essentially all the information needed to uncover the general law using mathematics. With some exceptions, processing the model into the general law is thus commonly the first step in interrogating *DE* models. What it takes will be a major theme in what follows next.

Arguably, change due to a process is counterpart to the derivative of the general rate law as given by the equation(s) of the *DE* model. The general rate law itself, then, should consist of expressions that upon derivatization *regenerate* the original *DE*. Getting from the model to the rate law by mathematical workup will *solve* the original *DE*. The end product is the (formal) *solution*, here the rate law that was generated by a *solution process* (*procedure*, *protocol*); however, that such an expression even exists—that it might be identified and in what way—is not to be taken for granted. It turns out that for various *DE*s, including models in this book, mathematical theory can establish whether a solution does in fact exist, possibly together with some other useful properties. Yet, even if a solution exists and can be identified, there is no guarantee that it will consist of elementary functions or others that are readily manageable. Altogether, obtaining a formal solution is highly desirable in developing a theory and may have advantages in actual computations. But, as will be seen later in this chapter and book, it is not always the most practical route to take, nor mandatory in every case. In fact, current modeling of cellular processes relies heavily on alternative methods—less than formal solutions (or approximations) that, nevertheless, succeed in simulating complex behaviors.

Only a small minority of real-life model *DE*s or others known in mathematics have formal solutions. These are sometimes referred to as *in closed form, exact,* and *analytic*—specific mathematical properties that a formal solution should have. The term *analytical* is also used more loosely in mathematics to differentiate expressions and arguments using mathematical symbols and the associated thinking from geometrical approaches, such as those used in creating the concept of a derivative above or from other graphical and numerical approaches; these will be taken up later in this book.

Pessimism aside, science and engineering make good use of the limited range of *DE*s that *can* be managed using formal methods. There are currently no general recipes for dealing with every solvable *DE*, not even for the favored but more restricted linear *DE*s. Cellular process modeling employs formally solvable *DE*s, as well, but as already noted, major kinetic models involve demanding mathematics and are open to formal methods only partially or essentially not at all. Even when modeling complex cellular processes employs alternative solution methods, these are expected to accomplish what formal methods would if they were applicable. Out of a vast *DE* solving methodology, what follows will thus begin with basic formal methods and move on to other procedures that will serve the purpose here.

Solving ordinary differential equations

In solving algebraic equations, the solution(s) that satisfy the equations are *numbers*, or *numeric values*. Solving *DE*s, on the other hand, commonly generates a mathematical *function* known as the *general solution*, the general rate law. Eventually aiming for numbers associated with process behavior in various conditions will require both further workup and additional data specific to each case. This will produce a so-called *particular solution* corresponding to the special rate law. Examples and more detail on both types of solutions will follow.

The overall path toward a general solution will depend on the particular makeup of the model's equation—the types of function(s), their derivatives and coefficients, the presence or absence of a nonhomogeneous term. Out of a large repertoire of *ODE* solving methods, what follows here will sample from those that may be of interest in modeling and in this book.

Taking integrals

Recall once again that taking a given *ODE*, say an original process model, to be the derivative of the general solution, the latter will be obtained formally by integrating that *ODE* to generate an (indefinite) integral of that equation.

The operation of integration being historically and logically closely associated with the solution process, it also has found its way into the terminology regardless of how it will be done. It is not uncommon to find synonyms such as *integrate* for the solution process, *integral* for a solution, and *integral curve* for its graphic representation.

ODE solutions at the most elementary level can be recognized simply by inspection and some help from basic calculus. For example, the solution of $y' = x$ is $y = x^2/2$—this, using the above common derivatization formula y $(x^n)' = nx^{n-1}$ and asking what would have been y given that $y' = x^1$.

The formal solution of this equation would indeed call for taking integrals. Given the right candidates, this is, in fact, the preferred and conceptually straightforward way to solve single *ODEs*. Consider the simplest case, Eq. (1.1a), describing induced protein synthesis. A general rate law, the general solution, can be obtained by taking integrals on both sides of the equation using the above formula. This will yield $P = vt + C$, where C is the integration constant that can assume any value. The constant is there for mathematical reasons, but it can be put to good use as a link to specific real-life conditions if needed for the particular solution. Here, for example, it can represent a protein that was already there at the beginning of the observation. Technically an initial value, this is essential when the formula aims for the total amount at a given time point.

Before proceeding further, a word of caution on using biological examples. To develop what follows next, textbooks commonly turn to radioactive decay as a real-world example—presumably a simple, independent process, as it is likely to be observed. Preferring here a cellular example, protein degradation can behave similarly, as depicted by Eqs. (1.1b) and (1.2). This, however, keeping in mind that protein degradation and synthesis (Eq. 1.1a) are complex processes that may behave differently, depending on the context.

Back to the main storyline, integrating Eq. (1.1b) is somewhat more complicated than Eq. (1.1a). However, this being a major prototype in cellular and other real-world modeling, it is worth a closer look and, for this book, a special level of detail. Turning to the same equation, reproduced as Eq. (1.10a), note that the term on the LHS, dP/dt, harbors both variables. Integration, however, is contingent on first *separating* the variables, a common first step in solving many other *DEs* as well. An option that can become problematic if not altogether impossible, *separability* should not be taken for granted. As simple an equation as $y'' + xy = 0$ for example is not separable. Here separating the variables is straightforward—all it takes is multiplying both sides of Eq. (1.10a) by dt and dividing by P to obtain Eq. (1.10b).

$$\frac{dP}{dt} = -\alpha P \qquad (1.10a)$$

$$\frac{dP}{P} = -\alpha \, dt \tag{1.10b}$$

Since from calculus the *LHS* of Eq. (1.10b) also equals $d\ln P$, the derivative of $\ln P$ (where ln is the natural logarithm), integrating will lead to $\ln P = -\alpha t + C$, where C is the integration constant. Following with a few more standard mathematical steps will lead to Eq. (1.11a), where e is the base of the natural logarithms, the constant 2.718.

$$P = Ce^{-\alpha t} \tag{1.11a}$$

$$P = P_0 e^{-\alpha t} \tag{1.11b}$$

Interpreting Eq. (1.11a), protein degradation in that system would depend on the function $e^{-\alpha t}$, where α is a rate constant, a measure of functional unit activity, and again on an integration constant, C. In principle, the latter can again assume any value and is therefore also a system parameter, as defined above. Standing for all solutions, Eq. (1.11a) is thus the general solution of Eq. (1.10a).

To generate an actual numeric value for a given system, Eq. (1.11a) is used further to obtain a particular solution. Here again, the constant will correspond to the special case, say, intact protein still remaining after degradation of a certain amount of starting material, but without being replenished by ongoing synthesis (i.e., as given by Eq. 1.10b). Not surprisingly, computing the as yet unknown constant will need additional information. Substituting into Eq. (1.9a), it turns out that when $t = 0$, $e^{-\alpha t} = 1$ and, therefore, the initial concentration of the protein $P_0 = C$. In other words, the amount of intact protein remaining will be given by the particular solution, Eq. (1.11b), where the constant is taken to be the initial amount (concentration) of protein, a knowable quantity, as it was in the case of protein synthesis. Once again, the initial value is in its dual role, real world and mathematical, the latter a role which it will have throughout much in this book.

Higher order equations into latent power

Cellular modeling largely employs *ODEs* that are first order, that is, feature first derivatives only (Eqs. 1.10a,b, for instance), and to a lesser extent *ODEs* that are second order. *ODEs* with still increasing order may occur and are generally more difficult to manage. One way to cope, alluded to above, is to replace higher order equations with systems of first-order equations.

There is, however, another approach that aims to meet this problem head on. Valuable as a modeling tool, it is brought up here primarily to introduce an altogether different concept in solving *DEs*. The concept is significant largely as an opening to a major strategy in evaluating complex system behavior; that is, solving and analyzing systems of first-order linear *ODE*s, such as Eqs. (1.8) and (1.9a−d). Solving both higher order *ODE*s and first-order systems common in modeling also can be turned into solving an algebraic and much more convenient counterpart and identifying parameters that are a solution and have various other uses.

Consider, once again, the *n*th order general *ODE* prototype, Eq. (1.3), now with constant coefficients and without the nonhomogeneous term.

$$\frac{\frac{a_0 y + a_1 dy}{dx} + \cdots + \frac{a_{n-1} d^{n-1} y}{dx^{n-1}} + a_n \; d^n y}{dx^n = 0} \tag{1.12}$$

When solving equations conventionally, a starting equation is manipulated by, more or less, educated trial and error until a solution is found. An elementary example is given early in Chapter 2, Consensus/linear algebra. Here and sometimes elsewhere in mathematics, it is more practical and still legitimate to try and go the other way around. The end product is first anticipated from previous information and one then searches for a route from the starting equation. Turning to Eq. (1.12), experience and theory indicate that solutions most certainly will be exponential functions in the form $Ce^{\lambda x}$. Evidently in the same form as the RHS of Eqs. (1.11a,b) they have the coefficient α replaced by the parameter λ, a constant with a value that depends on the case, where x is the independent variable.

Assuming that the solution will in fact be in this form raises a point that is key to the entire approach. Recall from calculus that the consecutive derivatives of $e^{\lambda x}$, where λ is a constant, form a sequence $\lambda e^{\lambda x}$, $\lambda^2 e^{\lambda x}$, $\lambda^3 e^{\lambda x}$, ..., $\lambda^n e^{\lambda x}$. Whatever the derivative, the expression $e^{\lambda x}$, or $(exp)\lambda x$, is recurring throughout the sequence. As such, it is an *eigen* (German adjective for "own," or "of its own") *function*, also known as *latent*, *proper*, or *characteristic function*, of the sequence. Each multiplier, λ^n, that satisfies the sequence is an *eigenvalue*. Seen another way the sequence of derivatives is an example of an *operator* (see above) being applied to an eigenfunction, resulting in its being multiplied by an eigenvalue. Here applying the differential operator, dy, to differentiate the eigenfunction, $e^{\lambda x}$ will lead to $\lambda e^{\lambda x}$, the eigenfunction multiplied by the eigenvalue, λ, a constant. Repeating the operation will yield the sequence above.

As applied to higher order *ODE*s, straightforward manipulation of Eq. (1.12) (not shown) will convert it into a so-called *characteristic equation*. In this equation the derivatives are replaced by constants, λ^n, eigenvalues raised to a power that equals the order of the original derivative as seen next.

$$a_0 + a_1 \lambda + a_2 \lambda^2 + \cdots + a_{n-1} \lambda^{n-1} + a_n \lambda^n = 0 \tag{1.13}$$

In other words, by taking advantage of what may be called the *eigenproperties* of the anticipated solution, specifically that the derivatives of the expected exponential functions essentially differ only in their coefficients, λ^n, the eigenfunction, $e^{\lambda x}$, has been eliminated altogether from the solution process and what remains is an algebraic equation of the eigenvalues.

At this point still unknown, the eigenvalues are the roots, the values of λ that will satisfy the characteristic Eq. (1.12) and will be determined by conventional methods of algebra. Ultimately solutions of Eq. (1.12) will thus largely consist of expressions such as $C_1 exp(\lambda_1 x)$, $C_2 exp(\lambda_2 x)$, and so forth. The Cs are constants, as in Eq. (1.11a) and the actual values of the λs in successive terms may (among others) be distinct or repeating.

Laplace transform

Another approach to solving a variety of single and systems of *ODEs* as well as *PDEs* is useful in mathematics and ubiquitous in dealing with technological systems and control in later chapters. Among its major benefits—it can lead to particular solutions of a model directly given appropriate equations and the initial values. This includes nonhomogeneous *ODEs* that are not solvable by methods described so far and are common in technological modeling (see Chapter 4: Input into output/systems). Employing the Laplace transform, the solution process takes an algebraic route as well.

As implemented in a solution process each term on the RHS of a model *ODE*, say the function $f(x)$, will be multiplied first by the expression e^{-sx} and then integrated. This will yield its Laplace transform, $F(s)$, shown in Eq. (1.14).

$$F(s) = \int_0^\infty e^{-sx} f(x) dx \qquad (1.14)$$

The sum of the Laplace transforms of all terms is, in turn, the transform of the solution. Applying the related (not shown) *inverse*, Laplace transformation next will lead to solution equations that again are function of x or t. Both forward and inverse transforms of frequently used mathematical functions are available in published tables.

Why the symbol $F(s)$ on the LHS? It turns out that the transform functions of each term do not anymore depend on the variables x or t as before but on the variable s, as indicated by the symbol. They are said to be in the Laplace or s *domain*, a term frequently found in technological literature. Here, however, this is of no consequence since s is a given number whose value has to satisfy certain mathematical (convergence) conditions, safely assumed to be met. Moreover, the variable s is never used directly in computing transform terms, their sums, or back inverses that are entered in the tables.

Another method, in a sense similar, may be applicable to certain linear, nonlinear and nonhomogeneous *DEs*, as well. It employs *integrating factors*, expressions that, when multiplying each term of an equation, opens the equation to further manipulations that will yield the desired integral.

Miscellaneous

Two more general points about *ODE* solutions are still noteworthy. (1) In addition to the solutions provided by a standard solution process there exist so-called *singular solutions* that seem unrelated. There is, for example, an *ODE* in the literature (not shown) that has one solution, $y = xy' + (y')^2$ as well as another, $y = -x^2/4$, that is not obtainable from the first by merely changing constants. Reckon that a model such as this could describe a process that, offhand, could work in two different ways. Or that these solutions could represent two completely different real-world systems that nevertheless follow the same laws of quantitative change. (2) Solving *ODE*s or other equations, meaningful solutions are sometimes accompanied by essentially meaningless, so-called *trivial* solutions, generally disregarded. For example, $c_1 y + c_2 y' = 0$ is a legitimate equation, but when c_1 and c_2 also equal zero it says no more than $0 + 0 = 0$, trivial indeed.

Alternative and supplementary methods

Aiming for accurate models of cellular and other real-world processes often leads to difficult-to-handle expressions, as already pointed out above. Challenges in modeling with *ODE*s can arise right from the beginning of a solution process. They could, for example, be due to the model including nonlinear expressions or others where integration can be a problem. Later on there may be expressions that are unwieldy and, as such, are difficult to compute. It is, thus, often unrealistic to expect a solution process, especially of customized models, that goes all the way to the point where it will yield actual process values. That is, the numbers a particular solution is expected to provide, say of the amount of nutrient induced protein synthesized by bacterial cells.

To manage, nevertheless, there are alternative methods that here will fall within two types of approaches. The first turns to alternative mathematical formats that in principle can duplicate or closely mimic the results offered by the original model and are workable throughout. Used independently, but also in conjunction with the first, the second approach aims directly for process values, but using essentially empirical procedures. Still mathematically sound, these can be more or less related to the original model, the point being that one way or another they deliver. Both types of approach play a significant role in modeling and will be covered in more detail in Chapter 3, Alternative infrastructure/series, numerical methods. Comments here are intended to provide the gist of what is often the bread-and-butter of cellular and other complex process modeling.

Series methods

Major alternative mathematics are based on an, offhand, nonintuitive, but well-established notion. It implies that *continuous* mathematical functions, such as the *DEs* in this chapter, have counterparts in particular *sequences* of *discrete* algebraic expressions. The sums of these sequences could, in principle, duplicate the values of the original model. The mathematics stipulates that the replacement of the original equation will be accurate when the sequence has an infinite number of terms, in which case the sum is known as a *series*. In practice the number of terms one uses is generally limited. Modeling with *ODEs* the series will be given by sequences of polynomial terms, ax^n. Used in truncated form, these expressions are in effect *(polynomial) approximations* and commonly known as such. While less accurate, they, nevertheless, offer the substantial benefit of replacing differential expressions of the model with more accommodating and computer-friendly algebraic expressions of the approximation.

Prominent in this approach and ubiquitous in modeling is the *Taylor series* approximation, kinetic modeling included. Using a known value of a particular process or its model, it allows to compute unknown values elsewhere. The Taylor series is helpful in a variety of ways, here notable is the common usage in handling nonlinear expressions. In what is sometimes referred to as *linearization* one aims to approximate a function's nonlinear behavior at a chosen point and its neighborhood with a linear alternative. A more detailed description of the Taylor series approximation and the associated equation is provided in Chapter 3, Alternative infrastructure/series, numerical methods. However, the three or even two initial terms of Eq. (3.2) there, as given by Eq. (1.15) here, are often adequate.

$$y = f(x) = f(x_0) + f'(x_0)(x - x_0) + \frac{f''(x_0)(x - x_0)^2}{2!} \qquad (1.15)$$

$f(x)$ is here the as yet unknown (process) value, y, corresponding to point x, to be approximated by the Taylor series on the RHS. $f(x_0)$ is a known value at a neighborhood point, x_0, taken as reference, $(x - x_0)$ is the distance between both points on the x-axis, and $f'(x_0)$ and $f''(x_0)$ are the first and second derivatives at point x_0 of the function to be approximated. The Taylor series approximation is effective *in the small* (i.e., near the known point of reference, x_0) and less so with increasing distance. However, not every function can be approximated by a Taylor series. Those that can are sometimes referred to as *analytic*.

Ubiquitous and typically powerful over extended intervals or *in the large*, are *Fourier series* approximations. They are naturally related to *periodic* behavior, say of biological rhythms such as circadian behavior and have broad applications in electromagnetic technologies. Fourier series also can describe a wide range of *nonperiodic* curves including, for example, such with discontinuities and even such consisting of discrete steps. Available in various versions, Fourier series methods are extensively employed in mathematics including the solving and computing of DEs, notably *PDEs*.

Employing trigonometric functions such as sin and cos, the early terms of Fourier series in a most general form, will be:

$$\Phi_n(x) = a_0 + (a_1 \cos x + b_1 \sin x) + (a_2 \cos 2x + b_2 \sin 2x) + \cdots \quad (1.16)$$

where $\Phi_n(x)$ is the value of the function using n terms of the series. a_0, a_1, b_1, and so forth, are coefficients that consist of integrals of the same trigonometric functions of equation variables as described in some detail in Chapter 3, Alternative infrastructure/series, numerical methods.

Empirical/numerical methods

Practical needs often outweigh the demands of rigor or accuracy. Real-life model *DEs* may not have a known formal solution to begin with. More regularly, especially studying complex processes, say, glycolysis or the control of the cell cycle, conceivably solvable models are liable to turn into expressions that are hard to work with or hard to compute and yield particular process values. For these there are practical computation methods that are based on general mathematical experience. Known as *numerical methods* or *analysis*, they are generally indispensable in cellular modeling. Numerical methods, in fact, are an old tradition, ancient Mesopotamians computing the value of $\sqrt{2}$ is a classic example.

Applications of numerical methods to *ODEs* are by now a versatile and highly sophisticated undertaking. Depending on the case, it is possible to compute values of modeling-related mathematical functions, their derivatives, and integrals. There are also procedures to handle *ODEs* as such. Described in more detail in Chapter 3, Alternative infrastructure/series, numerical methods, two approaches to solving *ODEs* numerically follow here.

Euler methods, time tested, the idea is to solve *ODEs* by replacing a formal solution process with an approximate curve obtained using a geometrical approach. Underlying is the familiar notion that an *ODE* is the derivative of its solution, at every stage of a process, at every point of its trajectory. The derivative at each point, as will be recalled, has a counterpart in the slope of the tangent to the *ODE* curve at the same point. Assuming, also, that the value of the tangent reasonably approximates that of the curve, Euler methods allow construction of the solution curve stepwise. In a basic version, described in some detail in Chapter 3, Alternative infrastructure/series, numerical methods, the x-axis will be divided into n points separated by the same distance, h. The y values of the solution will be determined at each point successively and connected into the solution curve, if needed. Here noteworthy is that the method can be represented by a concise recursion formula (see Eq. 1.24b and accompanying text) that can be readily programmed for machine computation.

$$y_{n+1} = y_n + hf(x_n, y_n) \quad y_0 = c \tag{1.17}$$

The LHS here stands for the computed new value of the solution (curve), y_{n+1}, as it depends on the RHS showing the most recent value of the solution, y_n, and the increment as given by h and the slope, the value of the model *ODE* at that point, $f(x_n, y_n)$. The initial value y_0 is a constant, c. Offhand, the smaller h, the shorter the segments, and the more accurate the approximation.

It so happens that this so-called *forward* Euler method is not highly accurate. To become acceptable for various applications it can be improved by further processing with another numerical method of which several are available. In that case, the first method is known as a *predictor* and the second as a *corrector*.

Runge−Kutta formulas, with detail also deferred to Chapter 3, Alternative infrastructure/series, numerical methods, are an important class of numerical methods for solving single as well as systems of *ODE*s. Compared to the not-so-accurate Euler methods, they improve the accuracy substantially by providing special error correction terms computed in a still enhancing, stepwise way. Coming in several versions, with *RK4* being the most popular, the Runge−Kutta methods are *predictor−corrector* methods. They are usable as such or modified further to reduce the extensive computations they can require.

Simulations

Altogether, with solution processes of cellular *DE* models largely turning to alternative methods, these models formally become simulations. Kinetic models of glycolysis in erythrocytes above and glycolytic oscillations later on are cases in point. However, the term simulation commonly is more often used in a wider sense in the sciences and technology. That is when formal or empirical expressions are used to mimic process behavior under a variety of conditions. As it were, cellular process simulations take it farther when they turn to system organizational features, possibly combined with quantitative methods, as will be seen in Chapter 8, Organized behavior/network descriptors.

Ways with partial differential equations

With *PDE*s and *ODE*s employing differential expressions, solution methods have naturally much in common. Formal solution processes depend on the makeup of the model—the particular functions, the order, degree, type of coefficients, the presence or absence of nonhomogeneous terms. Elementary-level handling of *PDE* building blocks will again employ the methods of calculus such as differentiation and integration, and so forth. In exceptional cases, and perhaps after some manipulation, directly integrating the terms will solve a *PDE*; however, the initial and crucial stages in solving *PDEs* as well as others later on are often considerably detailed, advanced, tied to the particular case, and largely less open to generalization compared to *ODEs*. This can be due to various factors. For one, *PDE* models employ multiple independent variables as well as second and higher derivatives instead of the single independent variable, first-order *ODEs* common so far. Boundary and initial values, here disregarded, are another highly significant factor. Reckon that these would be specified and processed separately for each variable; they can be in the form of constants, as with the *ODEs* above, but in many real-world models their values would vary as given by special equations. Taken together, the mathematics of the boundary conditions frequently dominate the choice of an overall solution route, the nature of the final outcome, and whether a solution will be unique. What follows here will relate to some of the more common and accessible features of solving linear *PDEs*.

The ordinary differential equation *route*

A major option in solving *PDEs* formally, every now and then designed into the model, centers on finding *ODE* counterparts that are manageable by standard methods. One may want to simplify original model equations initially, if possible. Sometimes physical considerations will allow to disregard independent variables. For example, when a process is at steady state a two-variable *PDE*, one being the time, would thus automatically reduce to a single-variable *ODE*.

Other initial processing may employ methods that are common elsewhere in mathematics. *Change of variables* (i.e., manipulating original variables) may allow to define new variables that group terms together first so that further workup in reduced numbers will lead to an *ODE* counterpart. For this and other purposes, the methodology may involve other manipulations, say, multiplying or dividing the terms of the original equation by an appropriate expression and then perhaps rearranging the transformed terms into a more favorable format. A related strategy that employs similar procedures calls for using first a *transformation* that will lead to a *change in coordinates*—a change in frame of reference and in units. Transforming from the common *Cartesian x, y, z* coordinates that specify the three distances from the origin to *polar* coordinates where each point assumes values given by *radiuses* and *angles* is a typical textbook example. Solving real-world *PDEs*, though, more often calls for other transformations.

An example of a standardized solution process with some generality is the method of *separation of variables*. What follows next in barebones outline will take some patience for detail, but no mathematics beyond that so far. The method is geared largely toward linear, second-order *PDE*s popular in describing nature and can be used with any one of the three Eqs. (1.7a–c) archetypes. Take for example an Eq. (1.7a) type diffusion equation. In this model, diffusion is represented by the change in concentration X of a diffusing substance as function of time, t, and position, x on the x-axis. In symbols, one has $\partial X/\partial t = D_x(\partial^2 X/\partial x^2)$, where D_x is the diffusion constant. Rearranging will yield $D_x(\partial^2 X/\partial x^2) - \partial X/\partial t = 0$, a common form. Readers will recognize here the reaction diffusion Eq. (1.4), but omitting the reaction term and slightly rearranged. Notice also that in this model diffusion is taken to proceed only along one dimension, the x-axis. This is the simplest, *one-dimensional* version of a diffusion equation. Such are more likely to be manageable mathematically and are often used in theoretical work. Solving more realistic versions in three dimensions can be problematic.

Obtaining a solution of the diffusion equation will provide a general law for the change in concentration with location and time. To proceed with the solution again requires separation of the variables as it was solving *ODE*s (1.10a,b). What was relatively simple in that case here takes a different and longer route that makes use of other arguments that were already encountered above, solving *ODE*s. The objective will be to find the solution of the diffusion equation in terms of positions, say of diffusing substance in a cell, and the time. Still unknown, the solution will once again be anticipated (see also Eq. 1.12), here in a most general, the bare-bones form, $u(x,t)$.

Now, by experience the solution process can be expected to go through the form $u(x,t) = X(x)T(t)$. Evidently the solution is now a product of two functions that represent the dependence on space, $X(x)$, and time, $T(t)$, separately. To continue, these, in turn, have to be separated completely. If the solution is indeed in the form above, then taking the derivatives of the diffusion equation one has $X''T - XT' = 0$. Adding XT on both sides of the equality and rearranging further gives $X''/X = T'/T$. The variables are now entirely separate and the ratios on both sides are equal and share the same constant, λ. Since now $X'' = \lambda X$ and $T' = \lambda T$, there will be two independent *ODE*s to solve: $X'' - \lambda X = 0$, and $T' - \lambda T = 0$. As previous and common usage of the symbol would suggest, λ is indeed an *eigenvalue* of the equation system.

Going further through the solution process will generate a solution in the form $u(x,t) = (A \cos \sqrt{\lambda}x + B \sin \sqrt{\lambda}x)\exp(-\lambda D_x t)$. What does this formula stand for? (1) The first expression in the first parenthesis depends on the variable x and represents solving the separate X''/X, position-related function. The exponential term next depends on the variable t and represents solving the T'/T, time-related expression above. (2) A and B are problem-specific constants related to the initial or boundary conditions.

Other partial differential equations *solution methods*

Integral transforms such as the Laplace transform, Eq. (1.14), and especially the related Fourier transform and the ubiquitous fast Fourier transform computer algorithm may be applicable to solving *PDE*s.

Solving *PDE*s in some cases can turn to an altogether different rationale and employ *variational* methods. Common in seeking optima for processes, these will be introduced in some detail in Chapter 6, Best choices/optimization. As applied to solving *PDE*s, rather than direct workup of the original equation one initially defines a set of likely solution candidates based on the makeup of the original *PDE*. The solution process will identify the candidate that is the closest to the presumed solution as judged by quantitative criteria, here notably the Euler−Lagrange condition (Chapter 6: Best choices/optimization, Eq. 6.8).

Also of note that: (1) *PDE*s are solvable formally in terms of elementary functions is not a foregone conclusion, as with *ODE*s. It is more likely with small number of independent variable *PDE*s such as those common in modeling here and elsewhere. (2) The solutions of *PDE*s prominent in science and technology, among others, are in the form of Fourier series. (3) Not surprisingly, modeling with *PDE*s often involves alternative methods—approximations, numerical methods, as well as simulations.

Divining systems of ordinary differential equations

Solving for collective behavior of contributing processes, the whole with the parts, the customized and structured modeling prototypes go in different ways.

Customized

Predominant in modeling ongoing cellular processes, as will be recalled, customized models commonly employ eclectic *ODE* systems and equations are solved individually. That makes it difficult to generalize here or go into great detail, mostly advanced. In general terms, dealing with small as well as older large models such as the Rapaport glycolysis models takes into account pathway properties and then follows mathematical opportunity to simplify and access desired information. The glycolysis model, for example, centered on manipulating and solving (concentration, rate) equations for particular intermediates and taking overall throughput of the pathway, the flux, to be given by the rate of the last step of the pathway, the production of lactate. Process values were obtained using a mix of methods that included specific nonlinear *DE* solving methods, series approximations, and numerical computations. Published details of these procedures are scarce as is still common in much of process modeling elsewhere. Computational practices are more transparent in some of the more recent literature, hopefully on the way to becoming the norm.

Contemporary models have available software packages geared to *ODE* systems of various kinds. Important, but not discussed here, are programs that are intended to maximize model quality, facilitate its handling, and check its adequacy. The computations commonly turn to numerical methods, often variants of Runge–Kutta and Euler procedures, mentioned above and in more detail in Chapter 3, Alternative infrastructure/series, numerical methods. Certain *ODEs* and their systems behave erratically in some of these routines and are known as *stiff* or *unstable*. To manage would require extremely small-step sizes, meaning expensive large numbers of consecutive computations. Problems can be addressed with software that checks for stiffness in advance and guides toward appropriate numeric routines. Whichever, without a formal solution, strictly speaking most customized models are simulations.

Structured systems

To be solved collectively and formally, the system of simultaneous model *ODEs* is again assembled bottom-up from equations for each subprocess, here in a particular linear format. However, the solution process is essentially top-down. How each subprocess behaves when combined with all the others is calculated from collective properties of the system determined first. To see how this might work, a small and more basic version of Eq. (1.8) will serve the purpose. In this example there are only two dependent variables, y_1 and y_2, and the time, t, is the independent variable. Its equations are once more in the favorite modeling format—linear, first order, constant coefficients (here a, b, c, d), and homogeneous as seen in equation system (1.18a) next:

$$\begin{aligned} y_1'(t) &= ay_1 + by_2 \quad y_1(t_0) \\ y_2'(t) &= cy_1 + dy_2 \quad y_1(t_0) \end{aligned} ; \tag{1.18a}$$

$$\begin{bmatrix} y_1' \\ y_2' \end{bmatrix} = \begin{bmatrix} a & b \\ c & d \end{bmatrix} \begin{bmatrix} y_1 \\ y_2 \end{bmatrix} \begin{bmatrix} y_1(t_0) \\ y_2(t_0) \end{bmatrix} \tag{1.18b}$$

Meeting also linear independence requirements, recall that to solve for n variables takes n equations, making the system n *dimensional*. With two variables, system (1.18a) is two-dimensional or *2D*. Providing, in addition, initial values (Eq. 1.18a, far right) guarantees unique (general) solutions in terms of elementary functions.

Experience and theory indicate that solutions of Eq. (1.18a)—essentially belonging to the Eq. (1.1b) prototype—would again be in the form of Eqs. (1.11a,b). Here, exponential functions such as $e^{-\alpha t}$ will have the exponent α replaced by an equation system counterpart, usually denoted λ.

The immediate problem is to determine λ since the time, t, is already a known. Taking it one step further, as with solving Eq. (1.1b), this will involve the coefficients a, b, c, and d.

Breaking the ground

To focus on the coefficients, as a first step they will be relocated and placed together in a separate unit by putting Eq. (1.18a) in the different format of Eq. (1.18b). Having the coefficients in a separate, purely algebraic system opens the way to further workup using the powerful methods of linear algebra. Here to begin with, one may wonder whether by changing the format one is not also changing the model itself or losing significant information. The answer is no on both counts. Why this is so, how and why linear algebra would work here, and the additional tools it offers are important topics in Chapter 2, Consensus/linear algebra. The focus here will be on basics in applications related to *ODE* systems.

In mathematics, numbers or functions arranged in a particular order are *elements* of units called *vectors*. Readers will be familiar from geometry with vectors that describe a line with a particular direction as specified by its values on the x, y, and z coordinates and given in the same order. Here, direction is not a factor but the order definitely is. In equation system (1.18b), the two coefficients a and b make for a *row*, or *r*-vector, y_1 and y_2 are a *column*, or *c*-vector, as do y'_1, and y'_2 on the *LHS* of the equation system. The rectangular array, the first object on the *RHS* that hosts together the coefficients, its elements, is a *matrix*. For certain purposes matrices can be viewed as consisting of row or column vectors. Both vectors and matrices of numbers and functions can be manipulated further as independent units, including addition and multiplication; that, however, using the special rules of linear algebra, as described in Chapter 2, Consensus/linear algebra.

These and other relationships between vectors and matrices can be described by what sometimes is referred to in the literature as *vector* or *matrix equations*. Vectors are often denoted by lowercase, bold, italicized letters, here **y'** and **y**. Ditto matrices, but using uppercase letters, here **A**. A compact and eminently convenient notation, system (1.18b) for example will be represented by the very simple matrix-vector equation.

$$\mathbf{y}'(t) = \mathbf{A}\mathbf{y} \qquad (1.19a)$$

$$\mathbf{s}'(t) = \mathbf{N}\mathbf{v} \qquad (1.19b)$$

Not surprisingly, the convenient format also finds its way into metabolic modeling. Eq. (1.19b) for example is a matrix-vector form of the stoichiometric model equation systems (1.9a−d). In this form the change in pathway member concentrations **s**, with time, t, on the LHS, equals a coefficient matrix, **N**, that represents the stoichiometries of each step, **n** (see also Eq. 2.5 and accompanying text) and the respective rates of the reactions, v.

If you noticed a discrepancy in notation between systems (1.9a–d) and Eq. (1.19b), it is there because they follow two different conventions. The parent systems (1.9a–d) represent substance concentrations by uppercase S, common in process descriptions. Using the bold vector-matrix symbols would imply that it is a matrix while in fact it is a column vector, counterpart to $y'(t)$ in Eq. (1.19a). Dealing here with mathematics, the lowercase symbol is appropriate.

Direct integration

Recall that solving the single *ODE* (1.1b), upon integration the rate parameter α ended as a coefficient of the time in the exponent, Eqs. (1.11a,b). When it comes to systems of the same *ODE*s, the same approach should work with the matrix of coefficients, A, instead. This type of solution, for example, will be encountered in basic linear systems and control theories as discussed in later chapters. Integrating Eq. (1.18b) one would obtain the following equation.

$$y = e^{At}y_0 \qquad (1.20)$$

where y_0 is the column vector of the initial values. That incidentally, the initial values appear on the right of the exponential, e^{At}, termed the so-called *matrix exponential*, and not on the left as in Eq. (1.11b) where they could be anywhere, is not accidental or trivial, but is due to the rules of linear algebra for dealing with terms in a particular order (see Chapter 2: Consensus/linear algebra). However, obtaining actual values from matrix exponentials can be problematic. In favorable cases there is a formal way that begins with the solution process described in the next section. For practical purposes, computer implemented numerical solution methods of the type described above and in Chapter 3, Alternative infrastructure/series, numerical methods, are preferred.

Prospecting for the latent

Formal solutions of structured *ODE* system process models commonly take a different route. The overall objective of the solution process is to generate rate parameters that will allow to determine the ongoing, local behavior of each subprocess separately. Collectively, the behavior of the parts will provide the ongoing behavior of the whole, the overall complex process. Some of these parameters, moreover, can also open the way to characterizing more general patterns of global, long range collective behavior and taken up by the qualitative theory of *ODE*s below. They are also applicable in the design of systems and their controls as discussed in Chapter 5, Managing processes/control.

How this solution process works will be given in outline form, deferring significant but more advanced detail to Chapter 2, Consensus/linear algebra.

Consider, once again, an *ODE* system whose members are first order, linear, with constant coefficients, and homogeneous. The solution process will again anticipate that each member *ODE* has exponential solutions in the mold of Eqs. (1.11a,b) solving Eq. (1.1b). The solution process will aim for rate parameters, counterpart to the exponent α in these equations. To solve for the rate parameters, the solution process will make use once again of *eigenproperties* already encountered in solving *ODE* (1.12), as well as *PDE*s, by separation of variables. Here in yet another version, many coefficient matrices, *A*, among other matrices, can be made to yield numbers, *(eigen)values*, λ, and *(eigen)vectors*, *q*, such that their product equals the product of the same eigenvectors but with the original matrix. In equation form one has $\lambda q = Aq$. Solving this equation will yield the values of the eigenvalues, λ, corresponding to each subprocess and counterparts of the exponent α in rate Eqs. (1.11a,b). Unlike solving for actual process rates, single equations such as (1.1b), solving systems also requires the eigenvectors, *q*. These are column vectors that have entries corresponding to all variables and are also computed from *A*. For each original *ODE* there will be an eigenvalue and a matching eigenvector. A mathematician might again see here an operator, now a matrix that *transforms* an eigenvector by attaching an eigenvalue, an action analogous to that of the differential operator on Eq. (1.12). How the eigenvalues and eigenvectors are actually computed by methods of linear algebra can be found in Chapter 2, Consensus/linear algebra.

Having gone through the solution process, as with solving single equations, will generate for each member equation the general solution, the general law for subprocess behavior. It will take the form of Eq. (1.21a). The collective behavior of the entire system will be given by the sum of these solutions in the form of Eq. (1.21b):

$$y_i(t) = \exp(\lambda_i t)q_i \qquad (1.21a)$$

$$\sum_1^n \exp(\lambda_i t)q_i \qquad (1.21b)$$

where *i* in Eq. (1.21a) is an index for the particular subprocesses and their respective variables and q_j is the corresponding column eigenvector. The subscripts and superscripts on the sum symbol, Eq. (1.21b) indicate that *i* can take any value up to *n*, the number of variables and equations. A particular solution of an *ODE* equation system will also include constants that represent the initial values (see Chapter 2: Consensus/linear algebra, Eq. 2.35). Altogether powerful, this approach to solving *ODE* systems is not without limitations. Depending on their particular makeup, certain matrices do not have computable eigenvalues and eigenvectors—generally an unpredictable property in real-world models.

More on eigenvalues and rates

Evidently the dynamics of processes described by Eqs. (1.21a,b) and the like are intimately related to system eigenvalues. In processes with multiple contributions, those associated with relatively large eigenvalues could dominate the collective rate. Eqs. (1.21a,b) are also similar in form to Eq. (1.11a) where α could be a chemical rate constant. Nevertheless, eigenvalues are not synonymous with rate constants. They are, in fact, difficult to assign a physical meaning in the present context, eigenvectors similarly. Reckon that process rates are obtainable from Eqs. (1.11a,b) based on α only, but here it takes both eigenvalues and eigenvectors. Consider them mathematical system parameters. Bridging these to chemical rate expressions is generally difficult and is a significant barrier to using eigenproperties rooted in modeling biochemical kinetics, ongoing cellular dynamics. This is especially important, considering that customized kinetic modeling also describes nonlinear processes nonlinearly. Nevertheless, eigenvalue methodology has its place in more specialized cellular modeling, as will be seen elsewhere in this book. It was, in fact, employed to characterize time scale hierarchies already in erythrocyte glycolysis, above. Related linear algebra methods are also important in computational procedures, with more in Chapter 3, Alternative infrastructure/series, numerical methods. Two additional properties of eigenvalues are still due a comment.

At the system level, the eigenvalue makeup affects not only process behavior, but also the mathematics. System (1.18b) type $n \times n$ matrix models, with n rows and n columns, can have n *distinct* eigenvalues—say 1, -8, 4, 7, 5, one corresponding to each of the original equations. Especially with larger systems, there may be several solutions with the same eigenvalue associated say 5, 3, 3, 3, 9. In this case the eigenvalues are *degenerate*. Further workup is commonly more straightforward when eigenvalues are distinct. This will also be generally the case in what follows.

Another mathematical property of eigenvalues will have a significant role in the qualitative theory of *ODEs*. Recall from algebra that numbers commonly in use here, in biology, and in everyday life are classified as *Real*. However, in many cases, for example, process models with negative feedback, equations also use the *square root* of *minus one*, $\sqrt{-1}$, and related numbers. These by definition are *complex* numbers and handled as a combination of *Re(al)* and *Im(aginary)* terms. Given by the general formula $\lambda = a_1 + ia_2$, where a_1 and a_2 are real numbers and i is the symbol for $\sqrt{-1}$ (*j* in the engineering literature), real-life eigenvalues may consist of either one or combination of both types of component. Note finally that while mathematical methods are certainly the mainstays in solving and evaluating *ODE* system models, certain cases allow for *empirical* evaluation. For example, as based on the physical behavior of electric circuits.

A perturbing real world

Models described so far typically represent an idealized picture of process behavior. In real life, processes are unlikely to duplicate model behavior faithfully, whether in cells, elsewhere in nature, or in the artificial world. Deviations, with potential biological or technological consequences, could affect a process at a particular instant, a local property, during an extended time period, a regional property, or over the entire course of the process, a global property. They could, for example, be due to change in the inner workings of a process, say, unexpected, modest accumulation of inhibitory substances, temporary or extended. There could well be discrepancies with the actual environment, say, in the pH or nutrient concentration. Human factors may be significant—the actual setup and subsequent behavior of many artificial processes, scientific experiments included, only approximate the original specifications. Much of the same can be said about real life, cellular, and higher level biology when circumstances change; *adaptation* may be an issue.

That drastic change in process settings could be of practical interest is obvious. However, relatively limited discrepancies with expected behavior, say, of no more than 10%−20% are much more likely. Would such change, in fact, affect a particular complex process or subprocess? If so, which will change? How much? When? Is change permanent? Reversible? If so, completely or partially? Might this change create extreme conditions that will threaten biological survival and competitiveness or technological safety? Will change become an economic burden? Obscure the true outcome of an experiment? There may be answers in knowing the range of potential process responses and whether they follow recognizable collective patterns. Would these be predictable from its model? Point toward improvements or failure of an application in genetic engineering? True, issues such as these are often out-of-sight in cellular or technological processes where they are regulated by built-in layers of control and compensation. It is, nevertheless, instructive to consider what the effects of internal and environmental change might be on inherent process behavior. Modeling the consequences is an original and still continuing motivation for the mathematical approaches on the agenda next.

Modeling an altered process calls for new rate laws. Solving model equations, when possible, would provide information on new process behavior. It is reasonable to assume that the new and the original are related. Before proceeding to explore how, a word first on how equations related in a different, previously encountered situation. Recall that in setting up simultaneous *ODE* system models, member equations are required to be consistent and linearly independent. Both are well-defined relationships that are readily testable by standard methods. The solutions of altered *ODE* systems interrelate as well. Collectively they hold information on perturbed process behavior that is here a main interest. Yet, the collective nature of *ODE* system solutions is not obvious from the equations and is not expressible in routine mathematical terms. Recognized in the 19th century by Lyapunov and Poincaré, they went beyond the mathematics of their day to pioneer a wide ranging discipline known as the *qualitative theory* of *ODE*s, the qualitative theory or qualitative approach from here on. The name notwithstanding, it is largely quantitative and rigorous.

A major focus of the qualitative approach to perturbed processes is on *stability*. Two stability-related themes will be taken up here. With some overlap, they differ in emphasis. The first addresses the balance between lower level contributions to complex processes when the internal and external settings change. There is here a technological incentive to reveal if, and under what conditions, there will be a stable balance. One way to answer is to study process response to systematic perturbation. Such analysis produces sets of *ODE* system solutions under different conditions that can provide specific answers. However, the main motivation for turning to these materials in this book is instructive. The qualitative approach also offers a general view of possible global behavior patterns, their variety, and the prominent role played in process behavior by the quantitative makeup of system and environment.

The second stability-related theme focuses on processes that are already stable at steady state and now face change in the environment. What options are there to respond? Will stability be maintained over time? How does it all relate to the quantitative makeup? Open to *ODE* system approaches, this will also tie in with another of the original agendas of the qualitative theory. Mathematical and facing difficult models, it aimed to develop methods for characterizing global dynamic properties, stability among others, but without directly solving the original equations. One of its successes, a Lyapunov method with diverse applications, concludes this section.

Before proceeding, it is noteworthy that modeling response to perturbation is a paradigm with diverse applications elsewhere. Theory provides blueprints for exploring metabolic control and for monitoring stability in numerical computations, among various uses. Checking behavior under perturbation sometimes is employed to test a model as such; for example, whether it is overly sensitive to real-life imperfections in monitoring, maintaining and reproducing physical conditions, to neglecting presumed minor factors or to computer error.

Varying systematically the internal and external makeup

Glycolysis, once again, has the common image of a steady housekeeping process. However, in the budding yeast *Saccharomyces cerevisiae*, under certain conditions glycolysis oscillates. Models account for some features of the oscillation, which is widely attributed to the kinetics of phosphofructokinase, a key enzyme in the pathway. Yet, no biological role has been established for this behavior. This raises questions such as, what would make a regularly stationary process like glycolysis behave unusually? Why is there even the option? What other potential behaviors might be available to the process, if any? Would these be interconvertible? How? What might be the effects of mutations? Changing the medium?

A role for system parameters

Concerns such as these and others could arise with cellular functions in general. Using process behavior over time as a measure of the underlying factors, the qualitative theory described next applies to linear and lineariz-able processes. To begin with, notice that models so far essentially represented three broad categories of real-life factors—the type of process, the level of its activity and the particular environment. Their modeling counterparts in equation systems (1.8) or (1.18a,b) and (1.22) would be the mathematical *functions* that relate the variables, the *coefficients*, and the *initial values*. In these terms, given the process, so is the function used in the model. The coefficients and the initial values, on the other hand, may change with the particular settings; they are *parameters* of the system. Obviously, process behavior is intimately linked to the mathematical functions that relate the participating factors, a central theme in *DE* theory and of much in this chapter so far. However, when it comes to collective global−system properties, dramatic effects can be associated with the parameters.

Looking into the role of the parameters, the simplest system of only two equations with two (dependent) variables is again handy. Already available in equation system (1.18a) the model is recycled here and becomes system (1.22).

$$
\begin{aligned}
y_1'(t) &= ay_1 + by_2 \quad y_1(t_0) \\
y_2'(t) &= cy_1 + dy_2 \quad y_2(t_0)
\end{aligned}
\tag{1.22}
$$

Using the behavior of system solutions to explore the role of the coefficients and the initial conditions, the analysis of system (1.22) indeed begins with solving for the variables, y_1 and y_2, once again employing eigen-values and eigenvectors.

Visual patterns

That change in the quantitative makeup of the coefficients, the parameters that stand for the level of subprocess activity, will affect overall process is expected. That changing in this model the initial values, say, the amount of a cellular precursor will affect the process is intuitive as well; however, both of these effects do not act independently—the response to change in the initial values strongly depends on the quantitative makeup of the coefficients. It is the *interplay* between the makeup of the coefficients and the initial values that matters here. The interplay is revealed when y_1 and y_2 are solved for in systems with different coefficient makeup and the initial values are varied systematically in each. It turns out that collectively the solutions for each coefficient makeup fall into particular patterns. These patterns are not apparent from the equation system and, as noted above, cannot be expressed with the usual mathematical symbols.

The solution collective patterns do emerge *plotting* the individual solution curves (strictly speaking, good Taylor approximations) together. As such these plots have technological applications. Usually advanced and not taken up here, these aim to characterize the response to parameter change at steady state, that is, when the ratio between subprocess contributions is stable. Among the more basic practical questions that arise, how large will possible deviations be? Could they destabilize a steady state? If so, could the process be self-correcting?

Here, however, a main interest is the striking diversity these plots reveal in complex process dynamics and how these can depend on process parameters. As will be seen, even a small, two-variable system, (1.22) will generate highly diverse patterns depending on the particular combination of coefficients and initial values. Now, depicting processes collectively has its practicalities. Representing on a two-dimensional page the two variables y_1 and y_2, each with an axis of its own is routine. Showing also the time, t, on the same page would make the detail here problematic. Since time moves on similarly and uniformly throughout both processes, it can be omitted altogether and replaced by the arrows that point in the direction of progress. Technically the plots are in a *phase plane*, the exact meaning of the term *phase* is here immaterial. Collectively the plots of the *ODE* system solutions make for a *phase portrait/diagram*. Mining these plots for information on overall process behaviors as well as more specific properties is known as *phase (plane) analysis*.

Examples of two-variable (or *2D*) phase portraits are shown in Fig. 1.3A−D. The names of the plots derive from the behavior of the Eq. (1.22) system solutions near the origin of the coordinates (the y_1 and y_2 axis in the figures), the area of the portrait that provides information on steady-state behavior. The arrows show whether or not when changing the initial conditions, the process will return to the steady state, and where it might go if not. As can be seen in the figures, depending on the parameters—the coefficients and the initial values—these could be a *proper node* (Fig. 1.3A and B), a *saddle point* (Fig. 1.3C) or, among various others, a *center* (Fig. 1.3D). A point where all trajectories converge at infinite time such as in Fig. 1.3A is an *attractor* of the system, a generic term loosely referring to a common final destination, of neighboring curves. As indicated

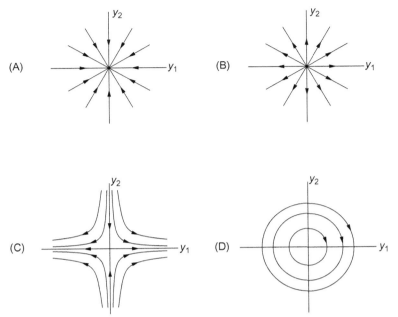

FIGURE 1.3 (A) Stable proper node, (B) unstable proper node, (C) saddle point, and (D) center.

by the arrows that point away or alternatively toward the origin, the node pattern in Fig. 1.3A is stable, in Fig. 1.3B it is unstable. The saddle point, Fig. 1.3C, is unstable while the center, Fig. 1.3D, is stable.

Quantitating the qualitative

How do phase portrait patterns correlate with the makeup of the coefficients? Visually diverse as the above phase portraits appear, they can, nevertheless, be traced back to the properties of only the two *eigenvalues* of equation system (1.22), λ_1, and λ_2. That refers to their relative magnitudes, as well as being either positive or negative, real or imaginary. This is not surprising since the eigenvalues find their way into rate formulas such as Eqs. (1.21a,b), where they are not simply multiplicands but powers of exponential functions and decisive in determining the course of the subprocesses. Fig. 1.3A thus represents a process with two eigenvalues both real and negative (λ_1, $\lambda_2 < 0$), and trajectories that converge on the origin making it an attractor of the stable system. A similar system but with real positive eigenvalues is shown in Fig. 1.3B with trajectories that now point away from the origin making it unstable. Fig. 1.3C, the saddle point, is associated with a real positive and a real negative eigenvalue. The center in Fig. 1.3D is its complex number counterpart, having one positive and one negative imaginary, but no real eigenvalues, and displaying the periodic character typical of imaginary eigenvalue components.

Altogether, *2D*, two-variable models are evidently instructive, aesthetically pleasing, and possibly useful. However: (1) *2D*, two-variable behaviors are difficult to generalize to three or higher dimensions (i.e., higher number of variables, say subprocesses). With more variables, system behavior commonly becomes qualitatively different, considerably more complex, and not obviously related or extrapolatable from lower dimensional analogs such as the *2D* system shown here. (2) Linear systems can exhibit periodic behavior, as seen above. That observed in cellular and higher level biological behavior is generally more complicated including nonlinear components and described by customized models. (3) Reckon that the dramatic effects on process behavior observed here are associated with the particular case in which process rate parameters are powers of exponential functions. There are, however, cellular and other biological systems, possibly such that obey different rate laws, where process rate parameter effects are notably minor and seem to be overshadowed by features of the organizational structure (see Chapter 8: Organized behavior/network descriptors).

Processes under perturbation

At steady state, in the literature often referred to as equilibrium, regular processes in cells or technological setups presumably function in a manner favorable to an entire cell or a production line. Phase plane analysis was concerned with whether or not once the initial conditions are perturbed a process will return to the steady state. Information on how this will happen and to what extent is indeed built into phase portraits, but the detail is not readily available from that format. Taking a direct look is the second theme of the qualitative theory on the agenda. Think of a chemical process represented by equation system (1.22) that is already running at a steady state. At some point the precursor concentration is moderately altered, counterpart to a perturbation of the initial conditions, the initial values of its rate equation. Will there be change in the output? If so, will the magnitude of a response relate to that of the perturbation? Would all or any particular subprocesses of a complex process be affected? Is change permanent or temporary? Once perturbed will the process remain within certain limits, perhaps moving to a different but similar steady state? Return to its original course? Partially? Completely? Get altogether off course? Similar questions arise perturbing other process parameters or when there is additional input during a process.

Stability options

In descriptive terms, when the response to perturbation is proportionate and the course of a perturbed process remains close to the main process the

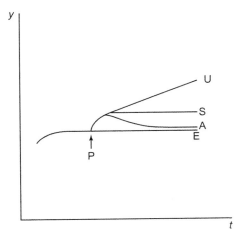

FIGURE 1.4   Stability behavior. Arrow—point of perturbation; U—unstable; S—stable; A—asymptotically stable.

perturbed process is *stable*. A perturbed system that eventually (almost completely) returns to the original steady state is *asymptotically* stable. If all perturbed trajectories of subprocesses in a complex process approach the equilibrium solution with time, the system is *globally asymptotically stable*. Systems that do not meet these criteria are *unstable*. Stability seen this way often correlates with how close a perturbed state will stay next to the equilibrium solution trajectory using various quantitative measures. In a stable system differences in values will be small say less than 10% compared to the main process and remain essentially constant while in an asymptotically stable system they will tend to zero as time approaches infinity. In an unstable system the difference is typically large and possibly growing with time.

Representative process responses to perturbing the initial conditions are shown in Fig. 1.4, where curve E represents the original equilibrium, curve A is asymptotically stable, curve S is stable and curve U is unstable. Note, incidentally, that linear processes typically have one steady state, but nonlinear processes may have several. In these cases, stability behaviors can differ between steady states, also between different stretches of the same process. A prominent role in current cellular process modeling, incidentally, is given to processes that can be stable at two separate levels. These, so-called *bistable* processes (see Chapter 4: Input into output/systems) are thought to represent cellular switches, say in triggering transitions between phases of the eukaryotic cell cycle.

Quantitative underpinnings

Two options, context dependent, will be considered here. Investigating perturbed *linear* processes is an established practice in engineering. It is also finding its way into biology and into other disciplines. Models indeed use system (1.22) prototype linear, first order, constant coefficient, homogeneous, initial value dependent equations. Focusing on the response to change in the initial conditions, but applicable also to change in other parameters, stability information is again available in the eigenvalues of the respective *ODE* systems. For example, a steady state is asymptotically stable when the eigenvalues are distinct and all have a negative real component. This makes good intuitive sense since it mandates that with time deviations from the equilibrium decrease toward zero. If one of the eigenvalues equals zero the system is stable by the definition above.

Local stability

Sometimes what matters is stability behavior of an equilibrium steady-state process at a particular point and its neighbors on the trajectory, rather than the global, long range stability properties described so far. It could be characterizing metabolic response to environmental change or evaluating stability of expressions that arise in numerical calculations, among others. With similar stability criteria and reflecting similar eigenvalue properties as those above, short-term models employ so-called *Jacobian matrices*. With wide applications elsewhere also for other purposes, Jacobian matrices here would represent how a complex process might respond to perturbing factors it depends on. In mathematical terms, a Jacobian matrix is one whose elements are functions, not numbers as they were so far.

As an example, assume a complex process that depends on two subprocesses, similar to system (1.22). In most general terms, the processes are function of two factors, x_1 and x_2, the independent variables, and are represented, say, by the formulas $f_1(x_1 + x_2)$ and $f_1(x_1 + x_2)$. The complex process would be described by a system of these two expressions. How perturbing the variables might affect local process behavior at equilibrium (steady state), $f(eq)$, is given by the partial derivatives of each of the processes with respect to the factors/variables. These are entered here as elements of a matrix as shown next where J is the symbol of a Jacobian matrix:

$$J = \begin{bmatrix} \dfrac{\partial f_1(eq)}{\partial x_1} & \dfrac{\partial f_1(eq)}{\partial x_2} \\ \dfrac{\partial f_2(eq)}{\partial x_1} & \dfrac{\partial f_2(eq)}{\partial x_2} \end{bmatrix} \tag{1.23}$$

Among the various uses of the Jacobian matrix one is especially notable. Without going into the detail, the rational for exploring process stability with the Jacobian matrix is that it is a linear approximation of presumed actual, possibly nonlinear process behavior that is reasonable as long as it is local. As it were, the elements of matrices such as (1.23) have counterparts in the Taylor series approximation, Eq. (1.15). They are analogs of the first derivative term there, the second on the RHS, and the first of the terms that would represent linearization. As Taylor series serve to linearize single nonlinear equations, Jacobian matrices are a widely applied matrix counterpart for locally linearizing nonlinear equation systems.

Bypassing a formal solution

Modeling nonlinear as well as other real-world processes being a persistent challenge, the qualitative approach is to explore alternatives. A way around to stability behavior is the second of the above two options. The *Lyapunov direct method*, or *second method*, is geared toward *ODE* models and still uses the stability criteria, but altogether bypasses an explicit solution. Its modeling range includes local, regional, and global stability properties of processes that can be linear and nonlinear, autonomous or not. In favorable cases, it may allow to go even further and solve equations and reconstruct entire phase portraits. Wide-scoped, contemporary Lyapunov direct method applications include cellular modeling and biological population dynamics as well as engineering, theoretical physics, and economics.

All of this, however, provided one that can first identify a special *Lyapunov function*, *V*. Technically this would allow to replace solving difficult model *ODE*s, essentially by integration, with more straightforward derivatization of Lyapunov functions. A derivative of the Lyapunov function (see below) is indeed an indicator of process stability at steady state. For example, when the derivative equals zero or is negative, an autonomous linear (equilibrium) system is stable. When it is strictly negative, it is asymptotically stable; globally so when all components of a complex process are asymptotically stable. A positive derivative indicates that the system is unstable. Working with nonlinear processes, the direct method is a versatile tool, among others in identifying when, over time, they will assume particular stability properties.

The detail of how the Lyapunov second method is used can be intricate. Here noteworthy is that: (1) Constructing a Lyapunov function, *V*, a special function is assigned to each variable. The latter can be those of the original equation, but often are redefined. (2) Appropriate functions can be identified is not guaranteed. When feasible, they may be motivated by analogous models of the energy properties of certain well-known physical processes. Significant especially in engineering, there are also more advanced identification methods when this is not feasible. (3) Lyapunov functions are constructed combining various terms. Terms taken together have to meet certain mathematical conditions. The *sum* of the separate terms *is* the Lyapunov function, V. (4) The derivative of V needed is that with respect to the system, dV/dy. It equals the sum of the values of each variable times the partial derivative of the Lyapunov function with respect to that variable. For example, with two variables, say y_1 and y_2, the derivative $dV/dy = y_1\partial V/\partial y_1 + y_2\partial V/\partial y_2$.

Being discrete/difference equations

Context

Much of what is known about cellular processes derives from processes in which change is gradual and apparently continuous, as described so far. These processes typically express the collective behavior of large numbers of discrete, but indistinguishable, functional units such as small molecules, virus particles, or whole cells in culture. Cells, in turn, also undergo major processes, mostly higher level, exhibiting change that appears stepwise and *discontinuous*. These processes typically take place in systems of discrete and *distinguishable* functional units that are present in relatively small numbers and can be *counted*. Examples are chromosomal mutations such as change in gene copy number and rearrangements, natural and induced change in chloroplast count, mitosis, and embryonic cell replication. In the artificial world discrete change might involve turning a page, manufacturing pieces of furniture, or the taking of the national census.

Discontinuous processes whose course is certain and not a matter of probability are naturally represented by the *discontinuous* or *discrete* difference in value between successive stages. Models of these processes can employ algebraic *(finite) difference* equations and are long established in modeling higher level biology and technological processes. They already have a history modeling cells as a whole replicating and undergoing other types of change, but applications to processes ongoing within cells are rare. At the same time, difference equations are a major tool in numerical methods for evaluating continuous models and in computer operations generally, for example, Eq. (1.17) and extended on in Chapter 3, Alternative infrastructure/series, numerical methods. The logic and the mathematical devices employed when modeling with difference equations often parallel continuous versions such as those described so far. Serving the purpose here will be some basic features of difference equations and their workings leading up an important application and to a popular example.

Building blocks

A culture starting with a single, say HeLa cell, and producing two progenies at each of five successive synchronous divisions, will (assuming negligible concurrent death) consist of 1, 2, 4, 8, 16, and 32 cells, respectively. In a model, these numbers can be represented by the dependent variable, y_n, and correspond to y_0, y_1, y_2, y_3, y_4, and y_5, respectively, where y_0 is the initial value of 1. The subscript n stands for the current number of divisions. As was the time, t, in Eqs. (1.1a,b) and (1.2), it is the counterpart of x, the independent variable, but now defined only at discrete intervals of one integral unit each (i.e., 1, 2, 3, 4, and 5).

Making a difference/elementary units of change

As with continuous models, local change is also at the center of discontinuous versions. As a basic unit of change there is the *(forward) difference* between two successive values, say of a process. Using again whole numbers, integers, as the independent variable, n, it is Δy_n in Eq. (1.24a):

$$\Delta y_n = y_{n+1} - y_n \qquad (1.24a)$$

$$\Delta y(x) = y(x + h) - y(x) \qquad (1.24b)$$

Generally, however, change in the independent variable is not restricted to integer values. In many real-world problems, it is represented as a *step*, also known as *increment*, or *interval*, usually designated h, whose values are chosen as needed. In this more general format, change can be given in the form of Eq. (1.24b).

Discrete equations/multiple faces

Difference equations, like DEs, can be set up in various ways. Taking differences—employing the *differencing* operator in operator language—the number sequence 1, 2, 4, 8, 16, 32, ... for example, obeys the initial value Eq. (1.25a), where counting n again starts at the second member (i.e., at $y_1 = 2$):

$$\Delta y_n - 2^{n-1} = 0 \quad y_0 = 1 \qquad (1.25a)$$

$$y_{n+1} = y_n + 2^{n-1} \quad y_0 = 1 \qquad (1.25b)$$

Especially significant is that Eq. (1.25a) can also be expressed in the rather different form of Eq. (1.25b) which turns out to be a major workhorse in computer operations. In this form the same number sequence is represented as a *recurrence relation* or *recursion formula* (see also Eq. 1.17). Given an initial value, y_0, and repeating or *iterating* the same mathematical step it is a formula for computing all consecutive terms in a standard way, a highly desirable property in machine computing. Every linear, constant coefficient, homogeneous, difference equation is expressible as a recurrence relation.

Difference equation can also take a more general form closer to that of the *DE*s earlier in this chapter. An *ordinary* difference equation, with one independent variable, corresponding to an *ODE* is for example:

$$f(x) = a_2 y(x + 2h) + a_1 y(x + h) + a_0 y(x) \tag{1.26}$$

where symbols are as before. The *order* of the equation is defined as the difference between the largest and smallest value of h, here 2. In line with its continuous counterparts, Eq. (1.26) is of *degree* one and thus linear. Here homogeneous, it could include also nonhomogeneous terms. Difference equations with multiple independent variables, counterparts to *PDE*s, are in use as well. Again, the format preferred in modeling when possible are difference equations that are first order, linear, constant coefficient, homogeneous, and linearly independent.

Formal solutions/process and product

Theory here aims to identify general rules as well as obtain numerical values for discrete system behavior. Recursion formulas indeed go a long way in machine computation; however, when n is large and the number of computer operations likely even larger, computer burden will sometimes be reduced significantly using formal solutions instead. These would be general expressions that can be applied directly to the particular case, as above. Again an extensive area, formal solution processes are commonly customized. Using familiar algebraic methods there are similarities to solving and using *ODE* solutions throughout. These include methods based on eigenproperties and usage of a *Z-transform*, a discrete analogue of the Laplace transform. Solutions also here can be exponential functions where the sum of all solutions is the general solution. Moreover, qualitative behavior may parallel that of *DE*s as well. Sometimes, discontinuous expressions are better dealt with using continuous counterparts. A familiar example is the use of the normal (Gaussian) distribution, a continuous function, to approximate and compute properties of discontinuous, binomial probability distributions.

Fibonacci

A classical manifestation of a discontinuous process in biology is a pattern observed at the cellular level that is particularly conspicuous in certain higher level systems. Obeying the discrete number sequence 0, 1, 1, 2, 3, 5, 8, 13, etc., and expressible as a recursion formula (see below) it was known in the East before, but was published in 13th century Europe by Fibonacci, whose name it bears. Among much else, the sequence describes the cell number in populations undergoing asymmetric binary cell division (cell fission into two daughter cells each with a different fate) known in yeast, insects, nematodes, and plants. Most familiar in the latter, the Fibonacci sequence is especially prominent in phyllotaxis, the leaf and flower arrangements around the stem. Although responsible mechanisms are as yet to be established, common patterns strongly suggest biologically significant similarities in the underlying processes. The Fibonacci sequence also describes various static as well as dynamic properties elsewhere in the real world. It has found its way also into wide ranging applications, notably computer algorithms and the analysis of stock pricing.

Turning to the mathematics of the Fibonacci sequence, the first two members are defined as $F_0 = 0$ and $F_1 = 1$, the symbol $F(ibonacci)$, replacing y above. Skipping both, each following number equals the sum of its two immediate predecessors. The quotient of the successive terms tends toward 1.618, …, the "Golden Ratio," a visually pleasing proportion in objects from nature, including various statistics of the DNA double helix, as well as in art and architecture. The corresponding recurrence relation is $F_n = F_{n-2} + F_{n-1}$ for $n > 1$. Solving this equation formally is more complicated than one would expect for such an innocent looking expression. Given here by Binet's formula, the solution is $F_n = [\varphi^n - (1 - \varphi)^n]/\sqrt{5}$ where $\varphi = (1 + \sqrt{5})/2$ is indeed the golden ratio.

Joining forces/hybrid and mixed representations

To conclude on a more general note, consider that models in this chapter largely employed equations of one particular type such as differential, algebraic or polynomial, or difference equations. Real-world modeling is flexible and mixes equations as needed. Customized kinetic modeling often combines differential rate equations with algebraic equilibrium expressions. Technological process descriptions associate differential state equations with algebraic output relationships that will be prominent in later chapters dealing with systems and control. When needed, processes can also be represented by *hybrid* equations that harbor different kinds of expressions. Among others these include *difference-DEs* and *stochastic-DEs* that incorporate recurrence relations or probabilistic expressions into *DEs*.

Keep in mind

Cellular processes, as we know them in situ or in vitro, mostly result from *continuous* collective activities of multiple functional unit systems, homogeneous and heterogeneous. Models can be based on quantitative *change* in process values to uncover rules for general process behavior and, most often, to determine absolute process values, therefrom or otherwise.

Kinetic process models, primarily molecular and neighboring level, regularly turn to DEs, *DEs*. Predominantly *time* dependent, coverage can extend to current status, behavior over extended time periods, and an entire course of a process—*local, regional* and *global* properties. Mathematically favored are *ODEs*, that are first order, linear, autonomous (coefficients are constants), and homogeneous, given *initial* value(s) and describing a steady state. To accommodate more detail, here likely time and space dependent with explicit *boundary conditions*, models can employ *PDEs*, notably first-order *advection* and second-order, *reaction diffusion* equations.

Modeling cellular complex process, for many purposes disregards subprocesses and employs single *DEs*. Aiming to reconstruct collective behavior from multiple lower level subprocesses, each is first assigned an independent *DE*. Together these make for a *system* of *DEs*. Using systems of *ODEs*, major cellular models are of two kinds. Predominant in representing ongoing processes, *customized* models are wide scoped and can combine linear and nonlinear *ODEs*, possibly also with algebraic or other equations. More permanent process features oriented, *structured* models center on predefined, linear, *ODE* systems of the mathematically favored and linear algebra—ready kind above, possibly already in the form of *vector-matrix* equations.

Discontinuous or *discrete* processes are typically evident in systems of countable and distinguishable functional units, likely cases at higher levels of cellular organization. Models employ algebraic *difference* equations, notably in the form of *recursion* formulas as well as other computer-friendly discrete apparatus with close parallels in continuous *DE* methods. Choosing any kind of cellular model, availability, and cost of computer power can be significant factors.

New process information can sometimes be obtained by simply inserting, say, subprocess rate data and using the model equation to compute a collective rate. More general in scope are formal *solution processes* that generate formulas for universal process behavior and use these for specific process values. The solutions will be mathematical functions, preferably *elementary*, that by differentiation will *regenerate* the original model equation(s) and their derivative status. Solving a single *ODE*, it may be *integrated* directly by standard methods or after some manipulation. The Laplace transform is another popular option. Solving *PDEs*, generally more demanding, often aims to identify and then solve *ODE* counterparts. Mainstays in solving discontinuous process models are the same recursion formulas now computer implemented.

Real-world models regularly include difficult-to-handle expressions, often the nonlinear. To manage, there are *approximations*, notably the *Taylor series*, and computer oriented *numerical* methods such as the *Runge—Kutta* algorithms. Combining partial formal solutions with numerical computations formally makes for *simulations*.

Solution processes for *ODE* system *customized* cellular models take into account, on a case-by-case basis, reaction and pathway properties, and then follow mathematical opportunity to simplify and access desired information. Traditional, this can call for standard integration methods as well as other established equation handling together with numerical calculations. State of the art modeling is computer implemented by standardized software packages that, among others, select favorable algorithms and commonly lead to numerical computations, notably by Runge—Kutta procedures.

Solving *structured ODE* system models using methods of *linear algebra* replaces differential with algebraic expressions, which is generally advantageous in handling *DE* models. It is also required in computer operations. Regrouping equation building blocks into *vectors* and *matrices*, there may be formal solutions by generating from the latter *eigenvalues* and *eigenvectors* (see Chapter 2: Consensus/linear algebra).

Global process behavior experiencing even modest *perturbations* may depart from its expected or regular course. Raising issues of *stability*, it can become a practical concern. Using models, analysis by the *qualitative theory of ODEs* takes into account the nature of the process (i.e., the mathematical function(s) relating the various factors) and the quantitative makeup of process associated parameters the coefficients, and of the initial values. Systematically perturbing the initial values and plotting the solutions of *2D* (two-variable) *ODE* systems reveals highly diverse visual patterns, *phase portraits*, that are typical of an interplay of the initial values with the makeup of the coefficients. *Phase plane analysis* of the visual patterns can uncover process tendency toward a steady state, an indicator of stability. As observed over time also with singly perturbed, steady-state processes, they can be *asymptotically stable*, *stable*, or *unstable*, stability properties that correlate with the makeup of model eigenvalues. Analysis of stability behavior may be possible circumventing a formal solution process by the *Lyapunov* second method, given a suitable Lyapunov function.

References

[1] Tyson JJ, Chen KC, Novak B. Sniffers, buzzers, toggles and blinkers: dynamics of regulatory and signaling pathways in the cell. Curr Opin Cell Biol 2003;15:221−31.

[2] Rapoport TA, Heinrich R, Rapoport SM. The regulatory principles of glycolysis in vivo and in vitro. Biochem J 1976;154:449−69 and references therein.

[3] Tyson J. Laboratory, <http://mpf.biol.vt.edu/lab_website/publications.php>.

[4] Recent textbooks include: a) DiStefano, J., Dynamic Systems Biology Modeling and Simulation (Amsterdam, Elsevier/Academic Press, 2014). Comprehensive, with engineering perspective. b) Klipp, E., Liebermeister, W., Wierling, C., Kowald, A., Lehrach, H., and. Herwig, R., Systems Biology (Weinheim, Wiley-VCH verlag, 2009) c) Helms V.: Principles of Computational Cell Biology, (Weinheim, Wiley-VCH verlag, 2008). d) Philips R., Konev J., Theriot J., and Garcia H.G. Physical Biology of the Cell, (New York Garland Science 2013.

Further reading

Brauer F, Nohel JA. The qualitative theory of ordinary differential equations. New York: WA Benjamin; 1969; Mineola: Dover Publications; 1989.

Hildebrand FB. Finite difference equations and simulations. Englewood Cliffs, NJ: Prentice-Hall; 1968.

Levy L, Lessman F. Finite difference equations. New York: The Macmillan Company; 1961. Dover Publications, 1992.

Morrison F. The art of modelling dynamic systems. New York: Wiley-Interscience; 1991. Mineola: Dover Publications; 2008.

Petrovskiĭ IG. Ordinary differential equations. In: Aleksandrov AD, Kolmogorov AN, Lavrent'ev MA, editors. Mathematics: its content, methods, and meaning. Moscow: Russian Academy of Science Press; 1956. English: Cambridge: MIT Press; 1963; Mineola: Dover Publications; 1999; Chapter V.

Pozrikidis C. Numerical computation in science and engineering. New York: Oxford University Press; 1998.

Tenenbaum M, Pollard H. Ordinary differential equations. New York: Harper and Row; 1963. Mineola: Dover Publications; 1985.

Zachmanoglou EC, Thoe DW. Introduction to partial differential equations. Baltimore, MD: The Williams and Wilkins Company; 1976. Mineola: Dover Publications; 1986.

Chapter 2

Consensus/linear algebra

Context

Major functions in nature and in the artificial world are realized by complex, heterogeneous systems and their processes. Replication, metabolism, and differentiation in living cells, self-correction in healthy ecosystems, data processing by computers, music made by an orchestra are all examples of this. In Chapter 1, Change/differential equations, complex processes as a whole were taken to express the collective outcome, a *consensus* if you will, of lower-level contributions whether active or inhibitory, originating in the system or in its environment.

Quantitative models of cellular complex processes intended, for example, to identify consensus, evaluate possible intervention or optimize performance, regularly turn to *systems* of ordinary differential equations (*ODE*s). Offhand, models belong to one of two major prototypes, as described in Chapter 1, Change/differential equations. Customized models predominate in representing ongoing cellular processes and are eclectic and wide ranging, but highly dependent on approximation by alternative methods. Structured models geared more toward dealing with permanent process features, center on predefined, linear, *ODE* systems and are open to formal solution methods. Both approaches use an infrastructure of the same type, albeit in different roles. *Linear algebra* methods are part and parcel of computing, especially of large, customized models in their later stages. Linear algebra is at the core of the structured models which rely on its methods from their very beginnings.

Cellular process models using linear algebra methods exist already for various features of enzyme as well as cellular pathway activities. There is a substantial modeling effort to benefit from the systematic ways linear algebra infrastructure offers to deal with higher-level processes, especially in cellular networks. Already contributing significantly, *flux balance analysis*, for example, offers procedures for optimizing cellular pathways for biomass production as well as characterizing effects of pathway member properties or external interventions. Moreover, linear algebra devices serve as highly versatile modeling tools in other approaches to cellular processes, as will be described in more detail in later chapters.

Collective Behavior in Systems Biology. DOI: https://doi.org/10.1016/B978-0-12-817128-8.00002-X

While the motivation for using linear algebra in modeling processes is largely mathematical, there is also a biologically significant aspect. Linear algebra infrastructure allows to represent explicitly process features that cannot be specified in the customized approach and are essential in modeling permanent system properties. It thus provides for representing systematically and keeping track of which subprocesses in a system interact, and in what way. In other words, linear algebra infrastructure offers ways to include in models *functional relationships* between system members. This is a first and often sufficient step in taking into account an overall system's *organizational structure*, which these relationships make together. Modeling organizational structure, a crucial factor in many cellular processes, will be covered in Chapter 8, Organized behavior/networks.

Not surprisingly, linear algebra methodology is already a mainstay in modeling complex processes elsewhere in the natural and artificial worlds, such as in the physical sciences, engineering, economics, and various social sciences, among others. It is, in fact, becoming central to modeling technological system processes and control with biological counterparts, and will be discussed up in Chapter 4, Systems/input into output, and Chapter 5, Managing processes/control.

Linear algebra methods were encountered in solving structured systems of *ODEs* in Chapter 1, Change/differential equations. The emphasis there was on *what* they can do in modeling processes. But, without the background provided in this chapter, much was left open, especially concerning *how* and *why* they work. Such explanations for a much wider range of methods will be presented in this chapter, which offers a basic survey of common linear algebra tools for managing equation systems, the original purpose of linear algebra. Largely geared toward the mathematics and intended for its purposes, these tools come with new rules and a style of their own. Perhaps esoteric at first sight, they are formal, but nevertheless, have a proven track record.

To begin with, there will be a general introduction of the basic structures that linear algebra employs and how they might work. Maintaining a special *format*, expression building blocks are housed in custom mathematical objects, *vectors* and *matrices*. These are units that can be handled independently, and they have a role in representing processes and possible interventions, as well as in managing the mathematics. Vectors and matrices can be manipulated using various *operations*. Some modify their internal structure, others combine them as intact modules. Matrices may also have *determinants*, which are values calculated in a special way that are useful primarily as diagnostics or as other defining features of equation system solution processes.

With linear algebra for *solving equation systems* as the destination, *Gaussian elimination* (*GE*) is a prominent application in solving simultaneous *algebraic* equations. A major methodology, in the process it also reveals key mathematical features of the equation systems. *Determinants*, again, have a historical role in solving algebraic and differential equation systems, but, as noted above, currently serve mainly in conjunction with other protocols. To close with solving *systems of ODEs*, the main focus will be on how the critical components already encountered and taken for granted in Chapter 1, Change/differential equations, *eigenvectors* and *eigenvalues*, can be obtained and then used until the system is actually solved.

All that said, the linear algebra covered in Chapter 1, Change/differential equations, assumed here to be familiar, should be sufficient for the main storyline of this book. This chapter is meant for readers interested in a better understanding of what linear algebra models represent and how this might relate to the mathematics. As such it should also offer a taste of what is a major methodology in the modeling of a wide range of real-world properties, both dynamic and static.

How equations change into modules

In preparing to solve simultaneous *ODE* systems using linear algebra, say equation system (1.8), the variables to be solved and their coefficients were initially separated and regrouped. This created highly compact modules that could be manipulated as independent units and represented by only a single-letter symbol. As a common first step in setting up models with simultaneous equations, it is often the format in which models are found to begin with throughout the literature. Why go this way? How will it work?

To begin in familiar territory, consider solving an arbitrary, simple, two-variable/two-equation system.

$$3x_1 + 2x_2 = 7 \tag{2.1a}$$

$$7x_1 - x_2 = 5 \tag{2.1b}$$

What values of the variables x_1 and x_2 will satisfy both equations together?

In elementary algebra, as readers will recall, it can be done by determining one variable at a time. Here one could first eliminate either one of the variables, say, x_2. Multiplying Eq. (2.1b) by 2, adding the result to Eq. (2.1a) so that terms with x_2 cancel out, will yield $17x_1 = 17$ so that $x_1 = 1$. Substitution of x_1 into either Eq. (2.1a) or Eq. (2.1b) will yield $x_2 = 2$ and the problem is solved. In principle, the method applies with any number of unknowns, eliminating variables consecutively. In practice, however, this solution process has various shortcomings. For one it is too laborious for large systems. Among others, representing interventions could be problematic as functional relationships between system members would not be explicit. Moreover the progress of an extended solution process, algebraic and differential, may require monitoring and/or meeting specific conditions.

Linear algebra methods provide effective answers for such needs, as will be seen, while the traditional process does not.

Getting organized

Bringing equation system (2.1a,b) into a linear algebra format, say in preparation for being solved, will separate the coefficients and the variables placing them into two types of mathematical objects as in Chapter 1, Change/differential equations:

$$\begin{bmatrix} 3 & 2 \\ 7 & -1 \end{bmatrix} \begin{bmatrix} x_1 \\ x_2 \end{bmatrix} = \begin{bmatrix} 7 \\ 5 \end{bmatrix}. \tag{2.2}$$

In this format, the coefficients in Eqs. (2.1a,b) still keep their positions, but those of the variables x_1 and x_2 have been shifted. Should the skeptical cell biologist worry about losing information originally in the model? Not necessarily if one is satisfied by being able to recover the original system. That this is so can be seen when—not by coincidence, it is the well thought through general practice—multiplying 3 by x_1 and 2 by x_2 and taking their sum to equal 7 will reproduce Eq. (2.1a). Similarly, 7 times x, plus -1 times x_2 equals 5 recovers Eq. (2.1b).

The square object in brackets first on the left hand side, LHS, of Eq. (2.2) is an array known as a *matrix*. The two other objects are *vectors*. Each such unit can be assigned single-letter symbols. The common notation adopted here employs *bold*, *italicized* characters, *uppercase* for matrices and *lowercase* for vectors. Together this will make for a compact, readily manageable representation, Eq. (2.3).

$$Ax = b \tag{2.3}$$

where A represents the matrix of coefficients that multiply the unknown variables in the vector x to satisfy the constants in vector b. Equations such as this are often referred to in the literature as matrix or vector equations (with more in Vectors, below). Compactness will evidently be maintained even when the number of *elements* or *entries* in either mathematical object is very large. What about these entities will be of interest here?

Vectors

Readers will recall from geometry that vectors are lines with a direction. For example an arbitrary line between points A(3,4) and B(6,9) (not shown), where the numbers in parenthesis are values on the x and y coordinates, are understood to be always given in that *order*. Here the term "vector" is used in a more general way, meaning a format which accommodates any kind of element as placed, indeed, in a particular order. It could be the same, all numbers position coordinates, or a heterogeneous set consisting of, say, (1) the name of a gene, (2) a status number, 0 or 1, indicating whether or not it is active, and (3) a rate formula for its transcription. As shown in Eq. (2.4), in linear algebra such vectors can be *horizontal* (called *rows*) as well as *vertical* (called *columns*) making for *r*-vectors and *c*-vectors, respectively. Elements are often denoted by the symbol a_i according to their place in the sequence, where i usually takes any number up to n, the total number of elements:

$$r = [\, a_1 \quad a_2 \quad a_3 \quad a_4 \,] \quad c = \begin{bmatrix} a_1 \\ a_2 \\ a_3 \\ a_4 \end{bmatrix} \tag{2.4}$$

Matrices

At the center of modeling complex processes and intimately related to vectors, matrices are as versatile in accommodating various types of elements—even vectors and matrices or possibly in multilevel, *nested* arrangements. In any event, to use the mathematics of linear algebra, matrices and vectors have to be linear (additive) expressions. That excludes as elements products such as variables squared, x^2, or multiplied by another variable (xy), and other expressions generating nonlinearity. Matrices here will generally consist of numbers and will meet linearity requirements.

Two versions of how elements can be represented are shown next, matrices A and N. Matrix A is in its most general, mathematical form and will be taken up first. Matrix N has real-life elements, here numbers that originate in a cellular model.

$$A = \{a_{ij}\} = \begin{bmatrix} a_{11} & a_{12} & a_{13} & a_{14} \\ a_{21} & a_{22} & a_{23} & a_{24} \\ a_{31} & a_{32} & a_{33} & a_{34} \\ a_{41} & a_{42} & a_{43} & a_{44} \end{bmatrix} \quad N = \begin{bmatrix} 1 & 0 & 0 & 0 \\ 0 & 1 & 0 & 0 \\ 0 & -1 & 1 & 0 \\ 0 & 0 & 0 & 1 \end{bmatrix} \tag{2.5}$$

Inspecting matrix A it can readily be seen as consisting of *rows* (horizontal) and *columns* (vertical), say r and c in *number* (symbols p and q, or m and n are also frequently used in the literature). Each row or column is a vector. It turns out that this view has major applications, some of which will be encountered below. The number of rows and columns, $r \times c$ (r by c) is the *order* of a matrix. As seen above, each matrix A element, a_{ij}, may be given an *address* where the subscripts i and j denote the number of the hosting row and column, respectively.

Square matrices

Modeling the real world as well as the mathematics give rise to matrices with various numbers of rows and columns, not necessarily equal. However, most relevant here, and often elsewhere, are the more special *square* matrices. Having an equal number of rows and columns, as do matrices A and N above, the square matrix is a major player in complex process modeling since that is the format favorable for solving systems of equations. Recall that it takes n equations (rows) to determine n unknowns (columns).

Whether one normally reads horizontally or vertically, there is something peculiar, namely significant information on processes is often found in matrices diagonally. That is, in the elements on the top left to bottom right *axis*, the so-called *(principal or major) diagonal*. Looking once again at the simultaneous *ODEs* of system (1.8), this was related directly to the real world. Modeling, for example, a cellular network or a biochemical pathway, parameters (e.g., rate expressions, stoichiometric coefficients, etc.) of the reaction of each functional unit by itself, say, an enzyme or a metabolite, have their natural place on the diagonal, address a_{ii}. When processes interact with one another, interactions will be represented by off-diagonal elements, a_{ij}. For instance, if a_{22} is the reaction rate of the second enzyme in a pathway, a_{24} may represent its activity as affected by the product of the fourth enzyme. And in more general terms, single elements here represent the activity of a simple process on its own. Rows would represent a subprocess in a complex system, possibly interacting also with other processes. The matrix stands for the activity of the entire complex process.

To see how matrices can be applied in cellular modeling and express functional relationships in a process, consider matrix N, a so-called stoichiometry matrix. Common, for example, in modeling metabolism, it will be useful later in this book (Chapter 6, Best choices/optimization). Matrix N is directly derived from the model of the first four steps of the glycolytic pathway given by Eq. (1.9) in Chapter 1, Change/differential equations. If that equation system was represented in matrix form, N would be the matrix of coefficients whose elements represent the stoichiometry, that is, the number of molecules participating in each step. The sign of each element stands for the functional relationship. That the forward reactions of the diagonal represent positive (plus sign implicit) input or output relationships along the pathway is straightforward; however, there is also the reverse reaction with a minus sign off the diagonal in row 3. Feedback providing elements further removed elsewhere could be similarly represented off the diagonal, as mentioned previously.

Diagonal matrices

In addition to its role in representing real-world processes, the matrix principal diagonal also has a role in linear algebra methodology. GE procedures are a case in point. Especially noteworthy here are matrices that host *only* diagonal elements, all the other addresses being empty (i.e., equal to zero). Known as *diagonal* matrices, D, a prototype is shown below. Its companion, the *identity* matrix, I, is a special and widely used variant that has all diagonal elements equal to one. Multiplication of any matrix by I as shown in Eq. (2.14) will leave it unchanged, hence the name and the symbol. Multiplication by the identity matrix is thus the matrix counterpart to multiplying by the number 1 in algebra. I has specific applications in matrix algebra, among others, in providing Eq. (2.30a) with a workable equivalent in Eq. (2.30b).

$$D = \begin{bmatrix} a_{11} & 0 & 0 & 0 \\ 0 & a_{22} & 0 & 0 \\ 0 & 0 & a_{33} & 0 \\ 0 & 0 & 0 & a_{44} \end{bmatrix} \quad I = \begin{bmatrix} 1 & 0 & 0 & 0 \\ 0 & 1 & 0 & 0 \\ 0 & 0 & 1 & 0 \\ 0 & 0 & 0 & 1 \end{bmatrix} \quad (2.6)$$

Not shown, but also in use modeling cellular processes, are matrices with only zeros, appropriately called *zero* or *null* matrices. In advanced applications matrices are converted by so-called diagonalization procedures to diagonal matrices with eigenvalues only, facilitating further operations. Calculating the eigenvalues as described later on and entering them on the diagonal is a simple example (see Eqs. 2.6 and 2.7, below).

Operating by the rules

Given a process with particular properties that are described by matrices and vectors, what will it take to obtain useful information? Suppose the process changes naturally or due to human factors, how can change be represented?

Cellular process models so far routinely employ arithmetic operations (i.e., addition, subtraction, multiplication, and division). For example, the total synthesis and degradation of protein, complex processes were described by an equation that used addition, subtraction, and multiplication as given by Eqs. (1.1 and 1.2). One would intuitively expect that representing the same processes, but with their subprocesses now using linear algebra methods, would still make use of the same mathematical operations. Or, to put it differently, that the operations used so far with single numbers and single-member equations, sometimes termed *scalar*, will apply also with systems of equations in matrix−vector form. This is, indeed, most generally the case. However, as already seen before, mathematical expressions collectively may follow rules of their own; therefore, closer scrutiny is called for. What follows next will show some of the more common basic operations with matrices and vectors in modeling the real world, as well as in handling the mathematics. To begin with matrices, complex objects that they are, they admit operations at various levels.

Single element

At the most elementary level, matrix and vector elements are open to common arithmetic operations (i.e., addition, subtraction, multiplication, and division). In fact, many operations at higher level eventually translate into operations between corresponding single elements. In a cellular context, single-element operations could represent, for example, modifying the activity of a single cellular component—possibly an enzyme or a signaling molecule, perhaps due to a mutation or an external agent.

Intramatrix

Largely used for mathematical purposes, major players in intramatrix operations are the manipulations of entire rows. Three simple *elementary matrix operations* are in wide use, including the solution of simultaneous equations, namely (1) interchanging rows, (2) multiplication or division by the same number of all the elements of an entire row, and (3) addition or subtraction from a row of another row or its multiples. In fact, operations (2) and (3) were used in solving the system of Eqs. (2.1a,b) traditionally. Although not to be used here, note that similar intramatrix operations can be applied to the columns, with similar associated mathematics.

A unique intramatrix operation, *transposition*, is common in more advanced linear algebra. In effect, it rotates the matrix elements around the diagonal of a square matrix. As a result, the rows and columns of the original matrix become the columns and rows of another, its *transpose*. By transposing matrices D and I in Eq. (2.5), being *symmetrical* with respect to the diagonal, they would remain unchanged. More commonly, however, A^T, the transpose of matrix A, will be a different entity even while the diagonal remains the same. The transpose of A in Eq. (2.5), retaining the original elements, is:

$$A^T = \{a_{ji}\} = \begin{bmatrix} a_{11} & a_{21} & a_{31} & a_{41} \\ a_{12} & a_{22} & a_{32} & a_{42} \\ a_{13} & a_{23} & a_{33} & a_{43} \\ a_{14} & a_{24} & a_{34} & a_{44} \end{bmatrix} \tag{2.7}$$

Vectors, by the same token, especially column vectors, are often given as their row transpose to save space on the page without affecting the algebra. For example, the first column vector, c of Eq. (2.4), occupies four rows, but the transpose, $[a_{11}, a_{12}, a_{13}, a_{14}]^T$, needs only one.

Also noteworthy, but skipping the detail, is that matrices can be *partitioned* into smaller sub- or *block* matrices. Dealing with complex systems, this simplifies representations and allows us to focus on selected subsystems.

There is a universal feature of matrices and vectors that can be a major factor in some of the more essential operations, but is conventionally not represented explicitly. For example, square matrices such as A in Eq. (2.5) on the page look symmetrical, except for their direction, rows, and columns, which appear indistinguishable. The same applies for vectors such as r and c in Eq. (2.4); however, preserving the original equation format, Eqs. (2.1a,b), it is implicit in routine matrix (not only square) and vector operations that elements are *added* within the (horizontal) *rows* of matrices and vectors, but *not* within the (vertical) columns of either. A row thus may have a sum, but a column cannot, the column in effect being a vestige of several different original system equations. In other words, there is built-in *asymmetry* in matrix and vector structure and related operations. It affects operations within matrices and vectors, some just described above. The asymmetry is also at the bottom of some of the unique and initially counterintuitive features of letting these entities interact as standalone units, and are discussed next.

Intermatrix or vector

Models of real-world, higher-level processes rely heavily on operations with entire vectors and matrices. In fact, already the basic linear algebra scheme of modeling processes with equation systems involves multiplying, as a whole, a matrix of coefficients by a vector of the variables, for example, equation system (2.2) and its compact analog Eq. (2.3) and the various *ODE* systems in Chapter 1, Change/differential equations. Going further and modeling, say, induced change in complex processes, it is important to keep in mind that in the handling of entire matrices and vectors, the actual operations involve the corresponding single elements. In other words, the model will directly describe change in all, or only some, particular subprocess components, as discussed above. One could, for example, use one matrix to represent the effect of temperature on a number of particular enzyme activities or those of other cellular activities and let it modify the matrix of coefficients of an original model of these activities. The choice of an operation will depend on the problem. For other modeling and mathematical needs there are various combinations, whether of vectors or matrices only, or mixed, as seen above. What follows will be a short survey of some of the basics this can involve. The main focus will be on matrix operations, discussed first, with comment on vector operations afterward.

Mergers

Vectors and matrices can be simply merged by adding the rows or columns of one to those of another. A frequent and often first step in solving simultaneous equations is the creation of *augmented* matrices. This typically involves merging an additional column to an existing matrix. In the case of Eq. (2.2), this would combine the matrix of coefficients from the LHS and the column of constants from the RHS as seen in:

$$A|b = \begin{bmatrix} 3 & 2 & 7 \\ 7 & -1 & 5 \end{bmatrix} \qquad (2.8)$$

Matrix addition/subtraction

Two matrices with the same number of elements in each row and column can be added or subtracted by the corresponding element. For example,

$$\begin{bmatrix} 2 & 7 \\ -11 & 9 \end{bmatrix} + \begin{bmatrix} 3 & -1 \\ 1 & 6 \end{bmatrix} = \begin{bmatrix} 5 & 6 \\ 10 & 15 \end{bmatrix} \qquad (2.9)$$

Note, incidentally, that two matrices are *equal* when the corresponding elements are identical.

Matrix multiplication

Addition of the same matrix element by element n times is equivalent to *multiplication* of the matrix by n, and, thus, essentially follows basic arithmetic.

$$3 \times \begin{bmatrix} 3 & -1 \\ 1 & 6 \end{bmatrix} = \begin{bmatrix} 9 & -3 \\ 3 & 18 \end{bmatrix} \qquad (2.10)$$

With two or more different matrix multiplicands the sequence of operations becomes more specialized as seen when multiplying a matrix and a vector in Eq. (2.2). To see how the rules are applied in general consider the multiplication of the two matrices, A, and B in Eq. (2.11):

$$A = \begin{bmatrix} 1 & 2 \\ 3 & 4 \end{bmatrix} \text{ and } B = \begin{bmatrix} 1 & 1 \\ 0 & 1 \end{bmatrix}, \text{ then}$$
$$AB = \begin{bmatrix} 1 & 2 \\ 3 & 4 \end{bmatrix} \begin{bmatrix} 1 & 1 \\ 0 & 1 \end{bmatrix} = \begin{bmatrix} 1 & 3 \\ 3 & 7 \end{bmatrix} \qquad (2.11)$$

In terms of matrix structure, the multiplication procedure can be seen as a combination of row and column operations. To compute the product matrix AB on the RHS, multiply elements in the first row of A in their order by elements in the first column of B in their order. Adding the two products will yield the first element in the top row of AB, repeat in the same order with the first row of A, but with the second column of B and obtain the second element, top row of AB. The same sequence of operations is then applied to the second row of A and the two columns of B.

Taking what are element level operations step-by-step, the process starts multiplying the *first* element of the *first* row of A by the *first* element of the *first* column of B and the *second* element of the first row of A by the *second* element of the first column of B. The sum of the two products is placed at the top left of AB, as:

$$1 \times 1 + 2 \times 0 = 1$$

Repeating the same sequence of operations with the two elements of the first row of A, but with the elements of the *second* column of B will yield the element at the top right of AB:

$$1 \times 1 + 2 \times 1 = 3$$

Readers may want to convince themselves that the elements in the bottom row of AB were calculated similarly from the *second* row of A and both columns of B.

Order of multiplication

Most significantly, multiplying the matrices of Eq. (2.11) in reverse order (but still starting from the left) yields a different result:

$$\boldsymbol{BA} = \begin{bmatrix} 4 & 6 \\ 3 & 4 \end{bmatrix} \tag{2.12}$$

In contrast to elementary arithmetic $\boldsymbol{BA} \neq \mathbf{AB}$, the order of multiplication here does matter and, in fact, is associated with special terminology. When A is on the LHS of the inequality it is *pre-* or *left-multiplied* by B; on the RHS it is *post-* or *right-multiplied* by B.

A major factor in using linear algebra (and counterintuitive), the role of multiplication order may be rationalized as recalling that intramatrix row and column operations are asymmetric. Here, Eq. (2.11) represents the multiplication of the rows of A by the columns of B, while in Eq. (2.12) the rows of B are multiplied by the columns of A. Typically, the sum of the elements on the diagonal (generally known as the *trace*) remains unchanged and equals 8 in Eqs. (2.11) and (2.12).

There are, however, matrix pairs, which are most desirable and often indispensable in mathematical operations, and whose products are identical, irrespective of the order of multiplication. These are said to *commute* or to be *commutative*.

Two equal-sized square matrices are always multipliable. Nonsquare matrices can be multiplied only when the number of columns in the first matrix equals the number of rows in the second. What takes place when this is not the case can be seen, for example, when trying to multiply a matrix A with four columns by a matrix B with two rows. The consequences are evident when trying to multiply only two representative vectors, namely matrix A top row, vector $[a_1, a_2, a_3, a_4]$ and what in a text is usually shown as a transpose, the matrix B vector $[b_1, b_2]^T$. As seen next, in calculating out this equation

$$[a_1, a_2, a_3, a_4]\begin{bmatrix} b_1 \\ b_2 \end{bmatrix} \tag{2.13}$$

Elements a_3 and a_4 would not be included and the operation is meaningless.

That multiplying by an identity matrix, I, as described above, in fact leaves a matrix unchanged can be seen in:

$$\begin{bmatrix} 3 & 4 \\ 5 & 6 \end{bmatrix} \times \begin{bmatrix} 1 & 0 \\ 0 & 1 \end{bmatrix} = \begin{bmatrix} 3 & 4 \\ 5 & 6 \end{bmatrix} \tag{2.14}$$

See also the matrix in Eq. (2.30b).

Matrix division via inversion

With matrix algebra counterparts available for the arithmetic operations of addition, subtraction, and multiplication the operation of division is an exception. Consider, once again, Eq. (2.3) where matrix A and vector b were the givens and vector x was the unknown to be calculated. By the logic of simple arithmetic, vector x should be obtainable dividing vector b by matrix A, which is obviously impossible.

There is a way around this. Recall that in arithmetic, division is the inverse operation of multiplication, and equivalent to multiplication by the inverse. For instance, since in $a \times 1/a = 1$ both terms on the LHS are inverses by definition, it follows that $b/a = b \times 1/a$. This idea is also workable with square matrices and in the case of Eq. (2.3) above

$$x = A^{-1}b \tag{2.15}$$

where A^{-1} is the *inverse* of A defined by $AA^{-1} = I$ (also $= A^{-1}A$). For example, if

$$A = \begin{bmatrix} 2 & 5 \\ 3 & 8 \end{bmatrix} \text{ then } A^{-1} = \begin{bmatrix} 8 & -5 \\ -3 & 2 \end{bmatrix} \text{ and } AA^{-1} = \begin{bmatrix} 1 & 0 \\ 0 & 1 \end{bmatrix} = I \tag{2.16}$$

In other words, dividing by a matrix is contingent on having it first undergo intramatrix inversion operations. When these are feasible, which is not always the case, the matrix has an inverse or is *invertible*.

As it were, being invertible, or *invertibility*, is also the first of a number of mathematical properties of square matrices such as linear independence and some others mentioned below (see the paragraphs just below fig. 2.16) that are associated with the original system of equations having a unique solution. These properties usually go together and finding one allows to infer the presence of the others. As for terminology, invertible matrices are designated *nonsingular* as those that are noninvertible are *singular*. Not coincidentally, the same terminology will be used with determinants, which is described in the next section.

Intervector operations

Addition and *subtraction* of row or column vectors are essentially straightforward single-element operations between counterpart members.

Ubiquitous and, indeed, a striking example of the built-in asymmetry is the *multiplication* of vectors. Multiplying a row by a column vector, as described above, will yield another vector as shown in:

$$[a_1, a_2, a_3] \begin{bmatrix} b_1 \\ b_2 \\ b_3 \end{bmatrix} = a_1 b_1 + a_2 b_2 + a_3 b_3 \tag{2.17}$$

On the other hand, reversing the order and multiplying a column by a row vector yields a matrix like the one shown here:

$$\begin{bmatrix} b_1 \\ b_2 \\ b_3 \end{bmatrix} [a_1, a_2, a_3] = \begin{bmatrix} b_1 a_1 + b_1 a_2 + b_1 a_3 \\ b_2 a_1 + b_2 a_2 + b_2 a_3 \\ b_3 a_1 + b_3 a_2 + b_3 a_3 \end{bmatrix} \tag{2.18}$$

Finally, note also that creating a real-world model, deciding, say, which of two matrices or vectors will be on the right or on the left will depend on what is being described. Further processing will follow the operating rules that apply.

A value for a matrix or determinants

That square matrices play a major role in solving equation systems has been seen in Chapter 1, Change/differential equations. Having a unique solution depends on these matrices having certain defining mathematical properties, such as having an inverse, has been intimated above. Checking for these properties, in turn, can serve as a solution process diagnostic.

Another defining matrix property in this category is the *determinant*, the product of a well-defined sequence of intramatrix operations. The outcome is a computed number, which is a value for the matrix. Determinants can have a dual role in solving equation systems. Like their parent matrices, they can be used directly in some solution processes. They can also provide information if a system can be solved to begin with and define conditions for getting there in others. Turning here, first, to applications as a diagnostic and with more later on, a matrix determinant is a diagnostic, depending on whether or not it equals *zero*. For example, an equation system will have a unique solution when it is *not* equal to zero.

What is a determinant of a matrix? Consider a two-dimensional square matrix consisting of the first two members of each row and column of matrix *A* in Eq. (2.5). The object on the LHS of Eq. (2.19) has the elements of this matrix, but is now housed in brackets without overhangs at the top and bottom. These brackets stand for the determinant of the matrix $|A|$ or *det* (A) and indicate that its value will be calculated as given by Eq. (2.19), RHS. That means multiplying the elements on the principal diagonal to obtain $a_{11} a_{22}$ and subtracting the product of the elements on the secondary or minor (top right to bottom left) diagonal $a_{12} a_{21}$, as shown. That difference makes for the determinant of a 2×2 matrix such as in:

$$\begin{vmatrix} a_{11} & a_{12} \\ a_{21} & a_{22} \end{vmatrix} = a_{11}a_{22} - a_{12}a_{21} \tag{2.19}$$

Increasing the number of elements complicates the calculations. It is evident already with a 3×3 determinant, shown here to provide some sense of the process. Allowed to use elements of any row or column for a so-called *expansion*, choosing those of the first row will lead to Eq. (2.20) and its directly computable RHS.

$$\begin{vmatrix} a_{11}a_{12}a_{13} \\ a_{21}a_{22}a_{23} \\ a_{31}a_{32}a_{33} \end{vmatrix} = a_{11}\begin{vmatrix} a_{22} & a_{23} \\ a_{32} & a_{33} \end{vmatrix} - a_{12}\begin{vmatrix} a_{21} & a_{23} \\ a_{31} & a_{33} \end{vmatrix} + a_{13}\begin{vmatrix} a_{21} & a_{22} \\ a_{31} & a_{32} \end{vmatrix} \tag{2.20}$$

In this case, each element of the first row is multiplied by the determinant of the elements remaining in the other rows and columns. These determinants are known as the *minors* of the element in front, their signs alternating between minus and plus. A minor together with its sign is a *cofactor*. Creating these and higher determinants has formal rules that can be found in more specialized sources.

How and why can determinants serve as diagnostics? The arguments underlying the main application are elementary, but somewhat detailed. Disinterested readers can skip Eqs. (2.21a,b) and (2.22a,b) and associated text. To proceed, consider an equation system such as system (2.1a,b), but now using symbols.

$$a_{11}x_1 + a_{12}x_2 = b_{11} \tag{2.21a}$$

$$a_{21}x_1 + a_{22}x_2 = b_{22} \tag{2.21b}$$

Solving this system, one can again turn to the simple notion (see Eq. 2.15) that if $ax = b$ then $x = b/a$. As can be seen, Eqs. (2.21a,b) are in the $ax = b$ format. Since the as and bs in Eqs. (2.21a,b) are numbers and not a matrix and a vector of numbers, as in Eq. (2.3), it will work. To begin solving system (2.21a,b) one would need to separate the variables x_1 and x_2. Using, here omitted, simple arithmetical manipulations, each variable ends up with an equation of its own:

$$(a_{11}a_{22} - a_{12}a_{21})\ x_1 = b_{11}a_{22} - b_{22}a_{12} \tag{2.22a}$$

$$(a_{11}a_{22} - a_{12}a_{21})\ x_2 = a_{11}b_{22} - a_{21}b_{11} \tag{2.22b}$$

Inspecting the expressions in parenthesis on the LHS of both equations, they are the same. Moreover, the same expressions would arise if one were to place Eqs. (2.22a,b) LHS coefficients as two rows of a 2×2 matrix as seen in Eq. (2.19), LHS, and then calculate the determinant as described there. To serve as diagnostics does not require going further and actually solving these equations as will be done using Eq. (2.27).

Determinants can provide diagnostics for solution processes as such, even when the solution process employs a different methodology altogether. That is, as noted above, because they define a precondition for a solution to exist. How? Recall once again that if $ax = b$ then $x = b/a$. Consider solving system (2.21a,b) equations, employing the latter formula where x is any one of the variables, b stands for the corresponding value on the RHS of Eqs. (2.22a,b), and a is its coefficient—determinant (2.19) written out on the RHS. The point is that because of the basic rule in arithmetic that prohibits dividing by zero, the formula $x = b/a$ will be meaningful only when the determinant a is *not* equal zero. The same holds for determinants of other sizes. It turns out that having a determinant other than zero is a property of invertible matrices and, in that case, determinant (2.19) is similarly denoted *nonsingular*. A determinant that equals zero is accordingly *singular*.

Now, according to Chapter 1, Change/differential equations, and as will be seen below in the section, Linear (in)dependence, that an equation system has a unique solution requires member equations to be linearly independent. This is not always easy to determine, but is confirmed when the determinant of the coefficient matrix is nonsingular. Though non-singularity is often a favored property, sometimes it is desirable to have a determinant that, in contrast, is definitely singular, for example, in obtaining eigenvalues for solving common modeling *ODE* systems later on.

Also noteworthy are the following. (1) Whether or not a determinant is singular (i.e., equals zero) is sometimes apparent already in its makeup. Among others, determinants will be singular when they harbor at least one row or column with all elements being zero, or when two rows or columns are equal. (2) Determinants are subject to operations in the style of intramatrix row operations, but of more limited scope. (3) Calculating determinants of real-life models, especially of larger equation systems, commonly employs computers. Machine error can be significant. When computed determinants happen to have a very small absolute value, it can be hard to distinguish whether the value is genuine or a machine artifact masking a true zero. There are, however, computer routines specially designed to minimize the problem. (4) As with having an inverse, matrices are often called singular or nonsingular, depending on whether or not they have a zero-valued determinant.

Solving equation systems

Having developed the machinery so far, how will it serve when it comes to actually solving equation systems? Examples of three different approaches will provide a taste for what is a major theme in cellular and real-world process modeling. However, a comment is due first.

Whether algebraic or differential, solving solitary equations requires that they meet existence, uniqueness, and term linear independence conditions, such as those in Chapter 1, Change/differential equations. The same is expected of each equation if it is to become a member of an equation system. However, collectively there is now a new layer of relationship between equations,. This raises similar issues, two being of immediate concern.

Consistency

Introduced as a requirement in Chapter 1, Change/differential equations, descriptively, why this may be so using numbers can be seen next. Recall that equation system (2.1a,b) had common solutions that were readily determined. This, however, cannot be taken for granted as, for example, with equation system (2.23a,b) where Eq. (2.23a) is a duplicate of Eq. (2.1a), but Eq. (2.23b) does not duplicate Eq. (2.1b).

$$3x_1 + 2x_2 = 7 \qquad (2.23a)$$

$$6x_1 + 4x_2 = 10 \qquad (2.23b)$$

Trying the standard solution method, successful with equation system (2.1a,b), will end with a meaningless bottom line that reads $0 = 4$ and lacks a variable candidate for a value. The respective curves of the two equations would be parallel and noncrossing. This system of equations, thus, has no common solutions, and its equations are said to be *inconsistent*. To have common solutions equations have to be consistent.

Linear (in)dependence

In addition to whether or not solutions to an equation system in fact *exist*, which is not the case with system (2.23a,b), obtaining *unique* solutions can still be an issue in cellular and other real-world modeling. Recall from Chapter 1, Change/differential equations, that given n equations needed to solve n variables there are, offhand, three possible outcomes: (1) there is a unique solution, obviously the preferred outcome; (2) there is no solution; and (3) there is an infinite number of solutions. Whichever is the case depends on how equations relate.

There is one unique solution when all of the n equations are *linearly independent*. Recall that *terms* of single *ODEs* in Chapter 1, Change/differential equations, were linearly independent when they were not similar enough to be interconverted by simply changing or attaching numerical constants. Similarly, *equations* that are members of an ensemble and represented by the *rows* of a matrix will be linearly independent when one row cannot be expressed in terms of one or a sum of the other rows, whether or not they were multiplied before by a constant. In other cases, the columns are the ones to be linearly independent, again by the same criteria.

Linear independence of the rows can be stated in quantitative terms. The measure used in modeling is often the *rank* of a matrix, denoted as *r*. Definable in different ways, here it will be the maximum number of linearly independent rows of the matrix of coefficients, indeed, representing the maximum number of linearly independent equations in the system. Now a system with n equations in n variables and rank r has $n - r + 1$ independent solutions. There is, thus, only one unique solution when the number of equations, n, equals the rank, r; this is another way of saying that all equations in the system are linearly independent. The rank in real-world modeling is often not obvious by just inspecting a matrix, especially when it is large. But it can be computed, for example, using GE. The rank of a matrix is then another one of the diagnostics for unique solutions, joining invertibility and having the value of the determinant.

Real-life models, however, may not meet conditions for a unique solution. Trying to make use of as much data as is available, models can end up with more equations than variables. They are thus *overdetermined*. Although some of the equations may be satisfied by a common solution this is unlikely for the entire ensemble, which generally will have no solution. On the other hand, a system is *underdetermined* when it has the right number of variables and equations, but includes equations that are linearly dependent. These systems will have an infinite number of solutions. Arguably, in these cases, while the total number of system equations was appropriate, they were redundant and there was insufficient information to choose between different options for a solution. Both types of models can, however, be dealt with. This may call for adjusting the model *itself* as well as for approximations by special mathematical methods, among others, using least square procedures discussed in Chapter 3, Alternative infrastructure/series, numerical methods.

Eliminating the variables the ordered way

A prototype apparently originating in Chinese antiquity and long established in solving linear algebraic equation systems is *GE*, now called after the distinguished 19th century mathematician. In response to various computational needs, it now has variants that generally are practical, can operate on a large scale, and are computer-friendly. In addition to solving equation systems, *GE* methods provide computing routines for matrix problems that arise in other works, notably, protocols for calculating matrix determinants, inverses, and rank. The *GE* process also offers diagnostics for questions of existence and uniqueness of solutions.

Underlying the *GE* approach is the familiar strategy of dealing with one unknown variable at a time, which was encountered in solving Eqs. (2.1a,b). Designed to eliminate successive variables systematically, it can be readily formulated as an *algorithm*, a set of predetermined mathematical steps that will lead to a solution as are common in computer operations. *GE* methods employ sequences of elementary intramatrix operations. Which of these and when they will be used is tailored to each particular case with specifics that are here dispensable.

The next example will highlight the main steps in the *GE* of an algebraic equation system (adapted with permission from Ref. [1]) given in matrix−vector form:

$$
\begin{bmatrix}
2x_1 & - 2x_2 & + x_3 & - 3x_4 \\
x_1 & - x_2 & + 3x_3 & - x_4 \\
- x_1 & - 2x_2 & + x_3 & + 2x_4 \\
3x_1 & + x_2 & - x_3 & - 2x_4
\end{bmatrix}
=
\begin{bmatrix}
2 \\
-2 \\
-6 \\
7
\end{bmatrix}
\tag{2.24}
$$

To begin with, the column vector of the constants from the RHS is initially added to the matrix of coefficients, the LHS, to generate the augmented matrix:

$$
\begin{bmatrix}
2 & -2 & 1 & -3 & 2 \\
1 & -1 & 3 & -1 & -2 \\
-1 & -2 & 1 & 2 & -6 \\
3 & 1 & -1 & -2 & 7
\end{bmatrix}
\tag{2.25}
$$

Applying a sequence of custom elementary row operations (not shown) will lead to a *row-reduced* matrix:

$$
\begin{bmatrix}
1 - 1 & 1/2 & -3/2 & 1 \\
0 & 1 & -1/2 & -1/6 & 5/3 \\
0 & 0 & 1 & 1/5 & -6/5 \\
0 & 0 & 0 & 1 & -1
\end{bmatrix}
\tag{2.26}
$$

Row-reduced matrices typically have rows with the number 1 leading other numbers. There may also be rows with only zeros (not shown) that would be at the bottom. Where present, the leading number 1 can be preceded by zeros, in which case the position of the leading 1 corresponds to the order of the row in the matrix, as seen above. Matrix (2.26) is said to be in *echelon form*. At this point, information is already on display as to whether there is a solution and what type to expect, if any.

No solution

It could happen that at the end of the *GE* process so far the last row, for example, is 0, 0, 0, 0, −1. This would imply that a number on the RHS equals the sum of zeros on the LHS before. Obviously a contradiction, it indicates the presence of inconsistent equations in the system.

Infinite number

When a consistent $n \times n$ reduced-row matrix harbors rows with nonzero elements less than n, the system has an infinite number of solutions.

Unique

There is one, unique solution if there are n nonzero rows in the reduced matrix (here 2.26), with last of these rows having the number 1 in position *two* from the RHS as shown above.

Solution values

To assign actual numbers to the variables, x_1 through x_4, and thus formally solve the problem, it is helpful to keep in mind the origins of the parent matrix (2.26). Recall that this was an augmented matrix in which the last column, here column 5 of matrix (2.25), represented the total value of each equation, originally placed on the RHS of Eq. (2.24). The values of each of the variables that satisfy the system and solve the problem are then calculated from the last two columns of the reduced matrix (2.26), beginning at the bottom. The number 1 in position four, row 4, represents the coefficient of x_4, and the number -1 to the right is its value. Row 3 means that $x_3 - 1/5 = -6/5$ yielding $x_3 = -1$. A similar calculation has $x_2 - (-1/2 - 1/6 = -2/3) = x_2 + 2/3 = 5/3$ so that $x_2 = 1$. In row one, the numbers after the leading 1 cancel so that $x_1 - 0 = 1$ (i.e., $x_1 = 1$). The rank of the parent matrix, incidentally, is given by the number of rows with at least one number. Here, it is indeed 4 as expected, obtaining a unique solution.

Employing determinants

Determinants have a long, but largely past, history in directly solving simultaneous linear equations, including applications in enzyme biochemistry. These days they may have a role in providing more efficient computational alternatives at particular stages of other solution processes.

A typical prototype in solving equation systems using determinants is the classical and once extensively used formula known as *Cramer's rule*. To see it at work we again use the two-equation system (2.21a,b) transformed into system (2.22a,b), now with the coefficients on the LHS already represented as determinants. As discussed earlier, dividing the RHS by the LHS of each of these equations, where the latter determinant is not equal to zero, will directly determine either of the unknowns, as shown next.

$$x_1 = \frac{\begin{bmatrix} b_{11} & a_{12} \\ b_{22} & a_{22} \end{bmatrix}}{\begin{bmatrix} a_{11} & a_{12} \\ a_{21} & a_{22} \end{bmatrix}} \quad x_2 = \frac{\begin{bmatrix} a_{11} & b_{11} \\ a_{21} & b_{22} \end{bmatrix}}{\begin{bmatrix} a_{11} & a_{12} \\ a_{21} & a_{22} \end{bmatrix}}. \quad (2.27)$$

This will solve the equation system yielding one unique solution.

Exploring the latent

Solving systems of linear *ODEs* such as those of structured, cellular models serves major purposes in cellular and real-world modeling in providing actual process values, as well as revealing behavior patterns of complex processes as seen in Chapter 1, Change/differential equations. Key to the solution process is the matrix—vector format and determining the coefficient matrix *A*, its *eigenvalues* and the *eigenvectors*, also known as *latent* or *characteristic* values and vectors. These were shown without explanation in Chapter 1, Change/differential equations, but their origins and how they are computed are not trivial. Mobilizing, also, materials from this chapter, highlights of how these parameters can be obtained will be presented here. It will also be an opportunity to follow the later stages of an *ODE* system solution process to see what an actual solution might look like. This should also provide a more general gist of how linear algebra methods may in effect be put to work. Abridged, the procedure for obtaining eigenvalues and eigenvectors and their employment in solving an *ODE* system described here still involves some detail, but the underlying argument is straightforward and within the scope of this book.

To begin with, a review of a few familiar basics. The single equation format favored in modeling process dynamics in Chapter 1, Change/differential equations, was an *ODE* that is first order, linear, has constant coefficients, and is homogeneous. The advantages of this format carry over when equations are combined into a system and it is popular, also, in vector—matrix format. Aiming for a unique solution would take *n* linearly independent equations for *n* variables. As with single equations, expected model solutions would be exponential functions, each associated with an eigenvalue and an eigenvector, as yet unknown. How will they be determined?

Computing these parameters combines three basic mathematical relationships: (1) the model, Eq. (2.28a), (2) the expected general form of the solution, Eq. (2.28b), and (3) the derivative of the needed exponential function, Eq. (2.28c), as shown next.

$$y'(t) = Ay \tag{2.28a}$$

$$y = e^{\lambda t}q \tag{2.28b}$$

$$(e^{\lambda t})' = \lambda e^{\lambda t} \tag{2.28c}$$

where *y* is the vector of the dependent variables of each of the subprocesses, *t* is the time, $y'(t)$ is the derivative of *y* with respect to time, a rate expression, *A* is the matrix of coefficients of the equation system, and λ and *q* are its eigenvalues and eigenvectors. Readers wishing to skip the details of the computation will find the formula for calculating eigenvalues and vectors in Eq. (2.31).

In a series of simple steps, Eqs. (2.28a,b,c) can be condensed into a single, workable equation. To begin with, $y'(t)$ on the LHS of Eq. (2.28a) is substituted with the derivative of (2.28b) RHS, namely $\lambda e^{\lambda t} q$ as computed from Eq. (2.28c). y on the RHS of Eq. (2.28a) is substituted with $e^{\lambda t} q$ from Eq. (2.28b). The new equation (not shown) will end with $e^{\lambda t}$ on both sides, in turn, canceling out to yield Eq. (2.29).

$$\lambda q = A q. \tag{2.29}$$

Evidently, all the components involved in the computation—the eigenvalues, λ, eigenvectors, q, and their source, matrix A—are present in Eq. (2.29). To continue with the computation, however, calls for separating the eigenvalues and eigenvectors first. Now, without getting into details, although there is a matrix on the RHS of Eq. (2.29) the expressions on both sides are vectors, nonetheless. Here, subtracting one vector from a second vector will yield yet another vector. Moreover, the vectors are equal; if they were numbers, subtracting one from the other would end in zero. Similarly, subtracting vector Aq on both sides of Eq. (2.29), a more formal way to perform the same operation, will lead to Eq. (2.30a) where 0 is known as a *null* vector (all elements equal 0).

$$\lambda q - Aq = 0. \tag{2.30a}$$

$$(\lambda I - A)q = (A - \lambda I)q = 0 \tag{2.30b}$$

Essentially rearranging Eq. (2.30a) by taking q outside the parenthesis will lead to Eq. (2.30b), shown here in the two forms it appears in the literature.

This is a critical point in the solution process. Eq. (2.30b) is satisfied only if the expression in parenthesis, $A - \lambda I$, a square matrix, has a determinant that, unlike what was seen above, does equal zero. In other words, that is a condition for having a solution and is often expressed in the form of the *characteristic* Eq. (2.31).

$$det(A - \lambda I) = 0 \tag{2.31}$$

The characteristic equation turns out to be the formula for the actual computation starting with the eigenvalues, λ. To proceed, determinant (2.31) is written out first (not shown) as was done with Eq. (2.19), yielding its RHS. This yields a so-called *characteristic polynomial*. Recall that polynomials, encountered in Chapter 1, Change/differential equations, are the sums of multiple terms of the form ax^n. Their *roots* are values of x when the sum of all the terms is taken to equal zero. The roots of the characteristic polynomial are the eigenvalues. For small system matrices the roots or eigenvalues are obtainable by standard algebra. Computing eigenvalues for large-system matrices commonly involves numerical method alternatives, an approach discussed in a different context in Chapter 1, Change/differential equations, and

Chapter 3, Alternative infrastructure/series, numerical methods, and here left to more specialized sources. Altogether, an $n \times n$ matrix may have up to n distinct eigenvalues. Recall, however, that the same eigenvalue may appear more than once in a solution, in which case it is *degenerate*.

Turning to the eigenvectors, q, these will be obtained subsequently by going back to Eq. (2.30b). With A, I, and 0 already given, reinserting the newly known eigenvalues, λ, will allow to solve the unknown eigenvectors. There will be one, unique eigenvector attached to each eigenvalue even when the latter is degenerate.

Following these procedures, for instance, matrix (2.32)

$$A = \begin{bmatrix} 1 & 4 \\ 9 & 1 \end{bmatrix} \tag{2.32}$$

has two eigenvalues, -5 and 7, and two respective eigenvectors, here conveniently shown as their transposes $[2, -3]^T$ and $[2,3]^T$.

With the eigenvalues and eigenvectors at hand, the way is open to solve the *ODE* system. Imagine matrix A, Eq. (2.32), being the coefficient matrix of a structured model two-equation *ODE* system with dependent variables y_1 and y_2 or in vector notation $[y_1, y_2]^T$, and t as independent variable. The original system would have been:

$$\begin{aligned} y_1(t)' &= y_1 + 4y_2 \quad y_1(t_0) \\ y_2(t)' &= 9y_1 + y_2 \quad y_2(x_0) \end{aligned} \tag{2.33}$$

where $y_1(t_0)$ and $y_2(t_0)$ are the initial conditions.

Recall that the *ODE* solution would be in the form of Eq. (2.28b). Fleshing it out by inserting the respective eigenvalue or eigenvector pairs, readers can convince themselves that the solutions of system (2.33) are:

$$y_1 = C_1 2e^{-5t} - C_2 3e^{7t} \quad y_2 = C_1 2e^{-5t} + C_2 3e^{7t} \tag{2.34}$$

where C_1 and C_2 are constants that depend on the initial conditions. As expected, the eigenvalues are in the exponents and eigenvectors are entered as the coefficients. Evidently, Eq. (2.34) provides formulas that allow to compute actual values, say, of an ongoing process.

Keep in mind

Managing equation systems, the linear algebra format can group large numbers of building blocks as *elements* or *entries* of a small number of compact objects, vectors, and matrices that can serve as independent units. *Vectors* are one-dimensional sets of elements in a particular *order* taking the form of *rows* (horizontal) or *columns* (vertical). *Matrices* are two-dimensional arrays of ordered elements that can be seen as consisting of row and column vectors. Preserving the original equation structure and critical in handling, vector and matrix elements are additive within rows, but not within columns. Especially notable are square, diagonal, identity, and inverse matrices as well as vector and matrix transposes.

For modeling the real world, as well as for mathematical purposes, vectors and matrices can be subject to operations at various levels. Bottom-level *single* elements of either object are subject to common arithmetic operations. Midlevel, *intramatrix* operations include *elementary matrix operations* whereby entire rows and columns can be relocated, added, and subtracted and rearranging elements by *transposition*. Top-tier, *intervector* and *intermatrix* operations with entire units, in effect, involve single elements and have counterparts in arithmetic operations. As entire units, vector and matrix *addition* and *subtraction* are the same as in arithmetic. *Multiplication* obeys specialized rules. Two matrices and/or vectors may be *pre-* or *postmultiplied*—two different options that may or may not yield the same product. Matrix *division* is indirect and can involve multiplication by a matrix *inverse*.

Square matrices provide unique solutions for systems of n *linearly independent* equations for n variables. Associated features include: (1) *invertibility*; (2) a *determinant*, a numerical value obtained by a defined sequence of intramatrix operations that is not equal zero; and (3) *rank*, the maximum number of *linearly independent* rows that equals n. Models with more than n equations are *overdetermined* and have no solution; those with rank less than n are *underdetermined* and have an infinite number of solutions. Both types are dealt with by special methods.

GE is a matrix protocol for solving simultaneous algebraic equations that uses elementary row operations. Eliminating variables sequentially, an initially augmented matrix is eventually transformed into a *row-reduced* equivalent form. Its elements allow to determine whether the system would have one unique solution, an infinite number of solutions, or no solutions. Values of the variables that satisfy a unique solution are obtainable by minor further processing.

Determinants can serve in solving linear equation systems, sometimes independently, but more likely in conjunction with other solution methods.

Eigenvalues and *eigenvectors* of coefficient matrices, when they exist, are key to formally solving systems of linear simultaneous *ODEs*. Anticipating a solution in exponential form, employing a sequence of linear algebra operations, leads to a *characteristic equation* from which the eigenvalues are directly computable using standard algebra. Back substitution of the eigenvalues into the characteristic equation yields the eigenvectors. Inserting eigenvalues and eigenvectors into a standard, exponential solution format provides the *ODE* system solutions and formulas to compute actual values.

Reference

[1]   Hamilton AG. Linear algebra. Cambridge: Cambridge University Press; 1989.

Further reading

Brauer F, Nohel JA. The qualitative theory of ordinary differential equations, an introduction. New York: W.A. Benjamin Inc; 1969. Mineola: Dover Publications; 1989.

Cullen CG. Matrices and linear transformations. Reading: Addison-Wesley; 1972. Mineola: Dover Publications; 1990.

Faddeev DK. Linear algebra. In: Aleksandrov AD, Kolmogorov AN, Lavrent'ev MA, editors. Mathematics: its content, methods, and meaning. Moscow: Russian Academy of Science Press; 1956. English: Cambridge: MIT Press; 1963, Mineola: Dover Publications; 1999 [chapter XVI].

Searle SR. Matrix algebra for the biological sciences. New York: John Wiley and Sons; 1966.

Chapter 3

Alternative infrastructure/series, numerical methods

Context

In the real world, a first choice may not be realistic, but there could still be a workable alternative. Cells short of glucose substitute with amino acids; human carcinogens are screened for by testing mutagenesis in bacteria; people who miss the train take the bus. Initial models of cellular processes employ equations that describe change in a system associated with the process. To obtain useful information on process properties generally takes a solution process as described in Chapter 1, Change/differential equations. As also noted there, however, with major cellular as well as other real-world process models, it is generally not feasible with standard equation solving methods. Alternative infrastructure can be effective and, in fact, is in common use. A preview of how it may play out was given in Chapter 1, Change/differential equations, and will be assumed here to be familiar. Bread and butter in cellular modeling, alternative infrastructure will be taken up in some detail in this chapter, emphasizing methods for ordinary differential equations (*ODEs*), mainstays of cellular modeling.

Recall to begin with that solving *ODEs* formally can be described as a two-stage process. In the first stage, change is used to obtain a general law, a *formula* that can determine what is changing, and in a variety of conditions. It could for example serve to predict the amount of induced protein synthesized or the expected amplification of a signal inside a cell. Practical and theoretical needs often call for actual numerical *values*; tangible *numbers* such as say the ongoing concentrations of a process intermediate in that particular case. These values will be computed from the formula given by the general law, also taking into account the associated conditions, in the second stage. Much of this book so far has been devoted to managing the first stage by formal methods. To introduce the underlying concepts, examples were chosen because they would succeed and also guarantee results in the second stage. As pointed out above, in modeling the real world, success is often elusive throughout.

Collective Behavior in Systems Biology. DOI: https://doi.org/10.1016/B978-0-12-817128-8.00003-1

Aiming for first-stage general solutions, already single (scalar) *ODE* models can be problematic. For example, original equations, nonlinear but also others, may have no formal solution or none in the favored form of elementary functions. Finding integrals for particular functions of a model, of the essence in a solution process, is another common problem. Second-stage issues arise, for example, in computing actual numbers from first-stage formulas using computers, possibly because of the nature of the expressions or the size of the model. Apropos, the later kinetic models of cellular processes these days can consist initially of the order of 100 equations; there can be 1000s in other real-world applications.

Demanding models, nevertheless, are regular fare and dealt with by turning to mathematical alternatives. Solution processes can begin formally and then switch to an alternative. In some cases the alternatives are advantageous right from the beginning. There are alternatives that are more-or-less related to an original model *DE*, while others differ considerably. All have in common that they are *approximations*, expressions that deviate, to a certain extent, from the original model and, thus, an *error* that generally can be estimated.

What alternative method might be practical? As with creating original models, no size fits all, and the options may vary. The choice will depend on the nature of the process, what already is known about it, the purposes of the model, and the tools available for a solution process. Choosing alternatives in solving an *ODE* or other process model, one would obviously want to know how close the alternative will be to the original and if likely error will be tolerable. It might take into consideration specific questions such as: Is the trajectory of the underlying process rich or poor in detail? Is it periodic? Does it call for representation by algebraic or trigonometric expressions (sin, cos, etc.)? Does the model aim for the fine detail or the overall features (i.e., process values at a particular stage) over a particular stretch, or its entire course? Can the method be taken all the way to provide a process value?

Ideally, alternative methods more or less *closely* reproduce the *entire* interval covered by an original model. There are indeed cases where the ideal is approached by a so-called *best* approximation. In general, however, this is rarely achieved. Certain alternatives work well with the detail (i.e., *in the small* or *locally*). Others are better in approximating overall features (i.e., *in the large* or *globally*). The mathematics of creating alternatives can be motivated analytically, considering what expressions will best replace those of the original model. Other methods are essentially geometric—one approximates an original equation by looking for an alternative way to specify the position of its curve. Some methods combine both approaches.

A longstanding tradition of formal alternatives is based on the notion that the values of *continuous* expressions, including *DE*s, can be obtained from *discontinuous* algebraic expressions, *polynomials*. Dealing with algebraic expressions, as will be recalled, is a substantial advantage both mathematically and using computers. Wide ranging, employing polynomials as alternatives will be approached here in different ways. One option turns to *series solutions* that replace the original model with a particular type of alternative expression. Typical examples are the *Taylor* and *Fourier* series approximations that were introduced briefly in Chapter 1, Change/differential equations, and here will be discussed in greater detail. A second type does not use an alternative formula to begin with but employs special polynomials chosen to deviate minimally from the original model. These so-called *Chebyshev best* approximations are generally more advanced, but are worthy of a comment. Offhand, all these methods are free to take any value. However desirable, they have their limitations and other options may have to be considered.

Generally referred to as *numerical methods/analysis* is an eclectic collection of practical methods that is extensively used to obtain numerical values for cellular and other real-world processes. Once again approximate, these are known by experience to generate realistic values one way or another. Numerical methods go back to Mesopotamian and Greek antiquity and by now play a major role in computer implemented calculations. Here to begin with and wide ranging will be *interpolation* and *extrapolation* methods. As the series methods above, they employ polynomials to compute numerical values, but require in addition that the approximation overlaps with the original model, at least at some of its values. Applied more specifically to differential equations, other numerical methods serve to compute building blocks such as derivatives and especially integrals of functions, as well as to solve entire equations and equation systems. By now an extensive and advanced practice, here to illustrate this will be elementary examples, largely relying on familiar geometry and algebra. A few other approximation procedures, incidentally, are scattered elsewhere in this book.

Machine computation has a substantial, sometimes essentially exclusive role in cellular modeling. While enhancing modeling capabilities, interfacing *DE* models with computer technology also involves a transition to an alternative mathematics, namely algebraic. Fidelity to the original and error of various types again become issues that need to be addressed. Reckon that these and the way computation is carried out and its associated technicalities can significantly affect the initial choice of models as well as their subsequent handling. As it were, the mathematics underlying machine computation of even large models, with modest adjustments, will be essentially still the same as that described here, emphasizing single equation models—better vehicles for bringing out and putting in context the underlying principles.

One might argue that using alternative methodology, generally approximate, altogether diminishes a model. Reckon, however, that as already seen in previous chapters and contrary to the popular image of mathematics as the paragon of precision, approximation is a way of life in its theory and applications. Turning here to alternative methods may be the only way to get any kind of result; it is also generally not the first encounter with compromise in a model's history. Creating an abstract representation, one already omits substantial detail. Even when initially included in a model, minor factors will often be neglected as modeling moves along. Using derivatives is a linearization of sorts and lessens fidelity to the real world further. Turning to alternatives thus adds one more layer to several already existing layers of uncertainty. Less idealized, the mathematics is, nonetheless, still sound. Moreover, there is a constant effort to compute and minimize *error*, deviations from an original model. Often, depending on the particular case, this is a significant aspect, but is here left to more specialized literature.

Value formally/series solutions

The polynomial way

Approximating *continuous* mathematical functions by *discontinuous* multiple-term expressions is an old practice. Perhaps odd, offhand. That it can be accurate, nonetheless, has been established rigorously by Weierstrass in the 19th century. It holds, among others, for *DE* building blocks such as derivatives and integrals of functions entire equations and their solutions, as well. In practical terms that can mean that the value of a function, or for that matter the solutions of model equations, can be obtained from the sums of discrete, linear, *progressions* or *sequences* of numbers. Mathematical theory shows that this is possible when the terms are infinite in number; this, however, provided the sequence of numbers *converges*. For example, the sum of the numbers 1, 1/2, 1/4, 1/8, and so forth converges to a *limit* that equals 2. Even with an infinite number of terms, sequences of numbers such as these thus have a definite sum called a *series*. Ditto sequences of certain functions; however, depending on the case, terms may converge only over a particular range of a sequence, the so-called *interval* or *radius* of convergence, or not at all. In any event, fast convergence (i.e., over a few early terms) can be essential. In practice, the number of terms one uses is generally limited and the value obtained is in effect an *approximation* and commonly known as such. In many cases the resulting *error* can be estimated quite accurately. Not to be addressed here, in what follows it will be assumed to be acceptable.

The sequences for modeling *ODE*s consist of *polynomial terms*, ax^n, introduced in Chapter 1, Change/differential equations, where a is a coefficient, x is, as will be seen, a broadly defined independent variable and n is an exponent. When they are infinite and converge this will be a polynomial sequence whose *sum* is called a *polynomial*. In practice, as will be seen, alternative methods of this origin again employ only a limited number of terms and are thus *polynomial approximations*. Whichever is used, the new format will come with several advantages: being linear even when the original model was not, replacing differential equation operations with arithmetic operations (addition, subtraction, etc.) and generally working with more user and computer-friendly mathematics.

A general and longstanding prototype for so-called sequence solutions, in effect polynomial approximations, are *power series* shown here in a general form:

$$a_0 + a_1 x + a_2 x^2 + \cdots + a_{n-1} x^{n-1} + a_n x^n \cdots \tag{3.1}$$

where n as the exponent is the *degree* (i.e., power) and as subscript an index. While the general case in Eq. (3.1) can accommodate an infinite number of terms, usage in practice is commonly in limited numbers, again, say, up to $a_n x^n$. Power series such as these can be used to solve directly certain *DEs*, but the approximations here, popular, will be more specialized.

Fine tuning with derivatives

A primary contender in approximating continuous functions with polynomials, *ODEs* and others is the *Taylor series*. It was encountered in Chapter 1, Change/differential equations, where its early terms served to *linearize* nonlinear expressions, often key to modeling cellular and other real-world processes. The Taylor series approximation has, in addition, various other applications. Here noteworthy are three points. (1) Obtaining integrals of expressions that arise in a solution process may be critical, but difficult or impossible. Recall that many mathematical functions have no known integral. Replacement with a Taylor series generally offers expressions that can be readily integrated. (2) Routine *ODE* building blocks, including the exponential, logarithmic, and trigonometric functions are not readily machine computable as such, but can become so as approximated by the algebraic Taylor series. (3) The Taylor series is a natural tool in approximating solutions of *ODEs* formulated as initial-value problems, the already familiar format central in modeling the real world and in this book.

The Taylor series allows to compute unknown values, in principle an entire curve, based on a particular value of a process or a model initially decided by the modeler. In what is also called an *expansion* around the known point, x_0, an anchor if you will, the Taylor series in a general form is given by:

$$
\begin{aligned}
y = f(x) = f(x_0) + f'(x_0)(x - x_0) \\
+ f''(x_0)(x - x_0)^2/2! + \cdots + f^n(x_0)(x - x_0)^n/n! + Rn(x)
\end{aligned}
\tag{3.2}
$$

The two first terms state that the approximated value, y, will be given as a function of x, or $f(x)$, as specified by the other terms. Beginning the approximation, $f(x_0)$ is the *known* value and the rest of the terms are successive *derivatives* at that point multiplied by $(x - x_0)^n$, the *distance* on the x-axis of the *approximated* point from the one known, x_0. As can also be seen, the distance, in turn, is raised to successive powers, as expected of the *power series* this indeed is the case. Last on the RHS, $Rn(x)$, is a *remainder* term that represents the difference between the approximation and the value of the original function, the *error* if you will. It commonly can be ignored. Although not visible everywhere above, terms of the approximation on the RHS except for the last have as a denominator $n!$— this being equal to 0! and 1! for the first two shown terms and by definition equal to 1, they were not entered. Now, the title of this section implies that derivatives of functions are instruments for fine tuning the Taylor approximation. They are indeed the unique and potentially problematic component because usage of the Taylor approximation depends on its having the needed derivatives, not always the case. But, as seen in Eq. (3.2), the series also depends on the distance between points raised to a power and the factorial in the numerator. Expanding the Taylor series about the origin of the coordinates, where $x_0 = 0$, leads to the *Maclaurin* series, a variant also used in applications.

Readers could wonder whether Eq. (3.2) with its derivatives, typical of differential equations, is actually an algebraic polynomial and related to that in Eq. (3.1) (disregarding the remainder term). In fact it is, since the independent variable here, that which may change, is $(x - x_0)$. The derivatives here are formulas to compute constant values and are here counterparts to the coefficients, a, in Eq. (3.1).

Evidently the Taylor series was derived analytically, that is using symbols and the associated mathematics. Note, however, that using only the two first terms one would get the same result as if one were to use simple geometry. Estimating the value at the next point, $f(x)$, as yet unknown, would then depend on: (1) the known, if you will, initial value $y(x_0)$; (2) the slope of the curve, computable as the derivative of the known function, $f(x_0)'$ corresponding to the known point, (x_0); and (3) the distance between x and x_0. They will be multiplied as seen in the second term on the RHS. As it were, this is a potent combination of factors that was already present in originally developing the concept of derivative (see Chapter 1: Change/differential equations) and can also serve to approximate entire solution curves of *ODE* models by Euler methods as will be seen later on. The next terms of the Taylor series, however, they are not open to a convenient geometrical analogy and are sometimes rationalized as a compensation for using a straight line as the slope in the second term.

When and where is the Taylor series appropriate? Given that it depends on the slope at the expansion point, it will be absolutely reliable when it approximates a curve with a constant slope, namely in the uninteresting case in which the curve is a straight line. The trajectories of real-life processes, however, often curve, and their slope changes, albeit gradually, as the process moves along. The Taylor series approximation is thus most accurate *in the small*, that is, near the known point of reference, x_0, and less so with increasing distance.

Recruiting sines, cosines, and their integrals

Not surprisingly, sometimes it is necessary to approximate a curve *in the large*, over an extended range or over its entire interval. With extensive applications also beyond approximation (and more below), the *Fourier* series may be the right choice. Using integrals as coefficients, it can be handy when obtaining the derivatives needed for Taylor approximations are problematic.

The basic-building blocks of the Fourier series are sine and cosine functions as shown in Eq. (3.3). As readers well know, these are common descriptors of periodic, oscillatory phenomena. The Fourier series thus is a natural match and, in fact, a common approximation. However, mathematics shows that the powers of the Fourier series can extend substantially beyond. That, because, in principle, *all* lines can be seen as if they combine lower-level periodic curves. The Fourier series, thus, can mimic a wide range of *ODE* and other continuous curves of various shapes, some discontinuities included. As can be seen in Fig. 3.1, these could, for example, be: (A) a natural sinusoid wave, (B) a straight line, (C) a step function/square wave with sharp discontinuities, and (D) a semicircle, among various others.

In the most general version Fourier series can be given by:

$$a_0 + (a_1 \cos x + b_1 \sin x) + (a_2 \cos 2x + b_2 \sin 2x)$$
$$+ \cdots + (a_n \cos nx + b_n \sin nx) \qquad (3.3)$$

As shown, there can be any number, n, of cosine and sine terms where $a_0, a_1, \ldots, a_n$, and $b_1, \ldots, b_n$ are coefficients such as those shown immediately below, and x is the independent variable, here in radians ($360°/2\pi$, c. $57.3°$). Fourier series may employ only sine or only cosine

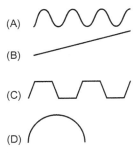

(A)

(B)

(C)

(D)

FIGURE 3.1   Fourier approximation.

terms and in limited numbers depending on the case. A polynomial approximation as well, it has to satisfy convergence and other conditions too. Used as an approximation, the equation might read as: $y = f(x) = a_0 + (a_1 \cos x + b1 \sin x). \ldots$

Like the coefficients of the Taylor series, the Fourier series coefficients, a_0, a_n, and b_n, are values given not directly by a number, but by a formula that employs functions to compute them. Here they are integrals of cosines and sines, as can be seen in:

$$a_0 = 1/2\pi \int_{-\pi}^{\pi} f(x)\,dx; \qquad a_k = 1/\pi \int_{-\pi}^{\pi} f(x)\cos kx\,dx;$$

$$b_k = 1/\pi \int_{-\pi}^{\pi} f(x)\sin kx\,dx \tag{3.4}$$

where $k = 1, 2, \ldots, n$, that is, the place of the respective kth term among the n total. The independent variable, x, is here allowed to vary as it would in a periodic description. In other words, between the limits of a cycle, 0 and 2π, more commonly represented as $-\pi$ to π. Although formulated for one cycle, the approximation can be used for intervals of any length. With minor adjustments of the coefficients it can apply, say, between arbitrary points a and b or even between minus and plus infinity, $-\infty$ and ∞.

Fourier series coefficients are often given in a more compact, complex number exponential version such as $1/2\pi \int_{-\pi}^{\pi} f(x)\,e^{-inx}dx$. By itself this would here be a minor point were it not that it also introduces one of the more acclaimed mathematical relations—that is, *Euler's formula* $\cos x + i \sin x = e^{ix}$ with i being $\sqrt{-1}$, the square root of -1. Using the formula here combines the cosine and sines in Eq. (3.4) into the single exponential term e^{-ixn} under the integral sign. It also is used extensively for other purposes in mathematics and the sciences.

In its variety of versions, the Fourier series has a major role in modeling applications other than those above. For one, as already seen in Chapter 1, Change/differential equations, in evaluating solutions of real-world partial differential equation (*PDE*) models where the series has originated (Fourier, early 19th century). Largely as an offshoot, the Fourier series is a major tool in a wide range of related physics and technology applications, X-ray crystallography among others. As it were, Fourier series methods also provide access directly to periodic components when present in real-world processes. An independent view of process behavior, a short detour from approximation practice, is worthwhile and indeed discussed next.

Innate periodicity

As it is in higher-level biology and elsewhere in the real world, periodic behavior of various kinds is well known at the cellular level—be it circadian rhythms, molecular feedback systems, cell migration, or others. Sometimes periodicity is evident but complex in appearance, especially when there are multiple contributions. Every so often, periodic components may not be immediately apparent altogether. To identify underlying components and evaluate their relative contributions, there are methods of *Fourier* analysis. An approach closely related to the Fourier series in its applications to cellular processes, it has already been instructive, for example, in characterizing the mobility and shape of entire cells, properties presumably associated with potential tumorigenesis, as well as in establishing periodicities in the molecular machineries of gene expression and cancer control.

Central to Fourier analysis is the Fourier *transform* (*FT*). Recall that in a regular periodic process the same time passes between the beginning and the end of each oscillation. This is its *time period*, *T*, and *1/T* is its *frequency*(ω). Each of the putative oscillatory components can have a frequency of its own. Using process data or a quantitative model, the separate oscillatory components, if present, can be seen by applying the FT of which there are different versions. In that here, data or model expressions will be multiplied by $e^{-i\omega t}$ and then integrated to provide the FT, $F(\omega)$:

$$F(\omega) = 1/(\sqrt{2}\,\pi) \int_{-.\infty}^{\infty} f(t)e^{-i\,\omega t}dt, \tag{3.5}$$

where t is again the time, ω the frequency, the integration is between times $-\infty$ and ∞, and the other symbols have their usual meanings. The FT is closely related to the Laplace transform of Chapter 1, Change/differential equations (Eq. 1.14), among other possibilities, in having an inverse and, usually, a reverse inverse that are used similarly. In addition to continuous *FT* versions such as Eq. (3.5), there is also a discrete version, *DFT*. Not shown here, it happens to be the origin of a highly efficient *fast Fourier transform* algorithm that is ubiquitous in machine computations of *FT*s.

In effect, an *FT* takes a model as it behaves in real time and converts it into a profile that represents the frequencies of the contributing oscillations and their relative contributions. In technical language, the original model is in the *time domain*, and the *FT* sends it into the *frequency domain*. In doing so it can document already well-known process components and perhaps also point toward so far unknown contributions. The frequency domain, as such, is a mainstay in modeling and developing applications in electromagnetic technology.

Quest for the best

Returning once again to approximation methods, given the shortcomings of the Taylor and Fourier series, among others, there has for some time been a search for ideal approximation methods. Such would be effective locally, in the small, as well as globally, in the large. A step forward that comes closer in both respects takes the form of so-called "best" polynomial approximations and was initiated by Chebyshev (pronounced Chebyshov) in the 19th century. If the Taylor and Fourier approximations will work best essentially where the model comes close to the approximation, the Chebyshev approach aims to fit the approximation closer to the model.

Underlying is the idea that certain mathematical functions can be approximated by expressions whose curves move regularly above and below that of the function to be approximated, crossing it back and forth. Illustrated by an arbitrary curve in Fig. 3.2, it makes sense because, offhand, there is no reason to assume that the approximation would move either above or below the target curve, only. What makes the actual approximating curve between these points "the best" is defined as *minimizing* the difference between its values at those points of the approximating curve that *deviate maximally* from those of the curve that is being approximated. The best approximation, incidentally, would be associated also with a best number of crossings. Various useful mathematical functions can thus be "best" approximated using *Chebyshev polynomials* (of the first kind). These can be defined as cosine functions or otherwise, as described in extensive literature. Best approximations may not be readily determined formally, but where available, can be calculated using computer-friendly recursion formulas. As it were, they also find other applications, including interpolation procedures, discussed below, as well as in evaluating derivatives and integrals of functions and in solving *PDEs*.

Another approach to approximation with similar motivation, and worth mentioning here briefly, is likely to be at least somewhat familiar. *Least square* methods, also polynomial, come in continuous and discontinuous

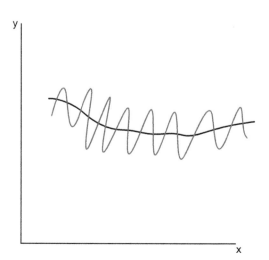

FIGURE 3.2 Chebyshev's best.

versions that aim to minimize the *square* of the *difference* in value between a curve and its approximation. Using values squared, one can combine values below and above the "true" value and compute an average of the latter for an entire interval of the curve. A popular application is in *curve fitting*, intending to provide an as yet unknown curve that will best describe, say, a set of data points. Here of interest is that least squares methods are also handy in approximating *overdetermined ODE* systems such as might arise in modeling cellular and other real-world processes, as discussed in Chapter 2, Consensus/linear algebra.

Note, finally, that both the Chebyshev and least squares approaches aim for a "best" outcome by seeking to minimize a difference between a desired model or curve, known or unknown, and another approximate, that can be computed. It turns out that similar strategies have been adopted elsewhere, for example, in the variational approaches to process optimization (see Chapter 6: Best choices/optimization).

Values the practical way: numerical methods

To obtain actual values for cellular and other real-world processes, modelers often turn to numerical methods. It could be an only, or only practical option. The formulas can be related to an original model but also differ. The point is that one way or another, for better or worse, they can actually produce a computed value. Numerical methods for *DEs* usually replace continuous differential with discrete algebraic expressions, generally an advantage when it comes to further handling and eventually using machine computation. There are procedures for calculating building blocks of original *DE* models as well as expressions that arise later during a solution process, not in the least for computing values for entire equations of change directly.

Not unexpectedly, numerical methods are usually approximate. *Error*, the deviation from a "true" target value due to shortcomings in duplicating the original model, can be a significant factor. It is thus common practice to evaluate error and, when beneficial, include in numerical formulas special terms intended to improve accuracy. The specifics of handling error generally vary with each case, call for more advanced and detailed handling, and will be of little concern here. In addition to built-in error such as this, *stiffness*, or *instability*—unreasonable error that can arise in using machine computation—is another and sometimes serious source of error, as will be discussed at the end of the chapter. Altogether, numerical methods today are an extensive, highly sophisticated methodology. Nevertheless, some of the underlying prototypes are based on simple, straightforward and, here, familiar notions. What follows is intended to provide a gist of what numerical methods are about.

Get in line

Given data on the rate of a particular cellular process at two temperatures, what will it be at some temperature in between? Knowing how long it takes to get from one place to another, when will you pass a particular point on the way? A table provides select values of a particular mathematical function, what will be an intermediate value? One way to meet such needs is to turn to *interpolation* methods and create a model which, offhand, is required to assume values *identical* to those already at hand, but could be approximate elsewhere. The term taken literally, interpolation methods are concerned with finding values *within* a known range, but the term can refer also to *extrapolation* when they are *outside* the range. Readers who, say, wanted to buy half a kilogram of an item priced at one dollar and expected to pay half as much were in fact using an interpolation method. Reckon that this assumed simple relationship is most likely described by a straight line. A numerical format of what, in effect, was *linear interpolation* will begin this section. Real-life processes are evidently varied and interpolation calls for more subtle procedures. Here to illustrate will be *Lagrange interpolation*, an example that at the same time will prepare for some of the more *DE* specific numerical methods later on.

   Linear interpolation is a most elementary method that, essentially as seen with the Taylor approximation, aims to determine unknown points on a curve using a given point as reference and a known or approximated slope. Turning to Fig. 3.3, consider the value of y of an internal point c lying between the two given extreme points a and b, the *interpolation points*. Lying on a straight line as shown in the figure, the value obviously equals the sum of the value of y at point a, and the increment in y from point a to

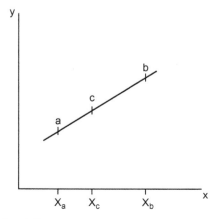

FIGURE 3.3   Linear interpolation.

point c. The latter is indeed proportional to the slope of the line as given by a straightforward ratio of the values of distances between two points.

Formally this becomes an *interpolation polynomial*, P:

$$P(x_c) = f(x_a) + (x_c - x_a) \cdot [(f(x_b) - f(x_a)))/(x_b - x_a)] \qquad (3.6)$$

Here $P(x_c)$ on the LHS is the desired value of the polynomial at the interpolated point c. On the RHS, $f(x_a)$ is the initial value of y at point a, the next term in parenthesis is the distance of the given points on the x-axis, and the expression in square brackets is the slope of the line between points a and b, the interpolation points.

Linear interpolation of this type can be extended also to complex curves using multiple interpolation points, however chosen. These points would be connected by straight lines and intermediate points would be determined locally by linear interpolation. Curves such as these serve also in other applications, for example, the trapezoidal method for numerical integration below.

Substantially more accurate and nuanced multiple interpolation point procedures are available and, in fact, provide a major approach to approximating mathematical functions. Assuming that the interpolation points represent an unknown mathematical function, interpolation aims to find a workable function approximating the values of the presumed original. Among others, *Lagrange* interpolation is well-established, often practical, and an opening to numerically computing derivatives and integrals (see below). The procedure again creates an interpolating polynomial, P_n, now a *Lagrange polynomial*, which satisfies n multiple interpolation points and can be used as a formula to evaluate points, as yet unknown, of the approximation. The resulting Lagrange polynomial can be relatively simple and user-friendly, but the way there is somewhat elaborate. Readers not interested in that detail are invited to skip this discussion and rejoin when it comes to the number and placement of interpolation points.

Given interpolation points, i, that belong to an as yet unknown interpolating function $y = f(x)$, the objective here is to find a Lagrange polynomial that can give the values, y, of this function at any point, x, between the interpolation points. Creating the Lagrange polynomial is once again a two-stage process. Key to the procedure and obtained in the first stage is the *Lagrangian*, L_i. That means there is a Lagrangian for each interpolation point, a ratio of two expressions that is to be calculated one at a time. In the numerator the Lagrangian takes into consideration the distance on the x-axis from an unknown point to be estimated at x, to all interpolation points other than the one chosen. This expression, as can be seen in Eq. (3.7a), is divided by a denominator, an expression for the x-axis, distances between all interpolation points. The second-phase generates the Lagrange polynomial, $P_n(x)$ as the sum for all interpolation points of the products of the associated Lagrangian times that point's value on the y-axis as given by Eq. (3.7b).

The machinery in these equations may look formidable, but its output, nevertheless, is an ordinary expression. As seen in Eq. (3.7c), the RHS it is a workable polynomial that can be computed readily in interpolation procedures as well as used for other purposes.

$$L_i = \Pi(x - x_j)/\Pi(x_i - x_j) \tag{3.7a}$$

$$f(x) = P_n(x) = \sum_0^n L_i(x) f_i(x_i) \tag{3.7b}$$

$$f(x) = a_1 x^{n-1} + a_2 x^{n-2} + \cdots + a_{n-1} x + c \tag{3.7c}$$

In Eqs. (3.7a) and (3.7b) the subscript i denotes any one of the n interpolation points x_1, ..., x_n, and subscript j in Eq. (3.7a) stands for all interpolation points other than the particular point chosen, i, or $j \neq i$. The Lagrangian corresponding to point i, L_i, is the ratio of the products, $\prod$ (Pi), of the various distance expressions delineated above. $\sum_0^n$ (sigma) in Eq. (3.7b) is the sum over all n interpolation points. The as and c in Eq. (3.7c) are constants, the superscripts are powers, and the subscripts are indices.

Offhand, one might assume that the greater the *number* of interpolation points used and their *placement* at equal distances (on the x-axis) will make for the most accurate estimate. Careful analysis, however, shows that this is often not the case and sometimes leads to significant computing artifacts. There is, thus, substantial theory on how to optimize these parameters. As it were, favorable interpolation point placements may be obtained from Chebyshev polynomials such as those mentioned above.

Interpolated curves sometimes turn out to be highly complex, possibly harboring discontinuities. There may be built-in computational issues due to too many interpolation points. In both cases, it can be advantageous to create piecewise, polynomial functions tailored to particular subintervals of the curve, *splines*. If needed, an entire interpolation curve can be constructed by assembling splines so that the values of their derivatives, that is, the slopes coincide where they join.

Calculating derivatives

Frequent *DE* building blocks, recall that the derivative of a function is counterpart to the slope of (the tangent to) its curve. In the most elementary of numerical procedures, the first derivative of a function, $f'(x_i)$, or dy/dx at point (x_i), is indeed obtained from the slope of the curve at that point. The slope, however, as is often the case also elsewhere, can be defined in more than one way, usually depending on the particular problem.

A *forward difference* formula for the segment starting at point (x_i) is shown in Eq. (3.8a):

$$f'(x_i) = (f(x_{i+1}) - f(x_i))/(x_{i+1} - x_i) \qquad (3.8a)$$

$$f'(x_i) = (f(x_i) - f(x_{i-1}))/(x_i - x_{i-1}). \qquad (3.8b)$$

Evidently the RHS of Eq. (3.8a) represents the slope of the approximating line, as did the expression in square brackets of Eq. (3.6). The *backward-difference* counterpart, computing in the opposite direction, is shown by Eq. (3.8b). Also in use are *centered-difference* formulas not shown. Without getting into detail, versions may differ with respect to the associated computation as well as the error, both potentially significant factors in actually choosing between options.

Methods for obtaining numerical derivatives more sophisticated than these are common. For example, extending on the similarity with linear interpolation, they can be based on multiple point expressions in the manner of Lagrange interpolation. In fact, finite-difference derivatives can be obtained by carrying out Lagrange interpolation first and directly differentiating the polynomial's terms next. A similar approach will be seen in numerical integration.

As for terminology, consider that the format of the numerical methods here is algebraic and discrete. The continuous derivative in Eqs. (3.8a,b) $f'(x_i)$, or dy/dx, is thus approximated by a *difference quotient*, $\Delta y/\Delta x$, its discrete counterpart here written out on the RHS of Eqs. (3.8a,b). Expressions such as these, as well as the more complex in other cases, for example those arising from Lagrange interpolation, are *finite-difference formulas*.

Computing integrals

Integrals of functions, readers will recall, play a major role in solving and evaluating *DE* models. Although many integrals are today available in tables, finding and computing integrals in specific real-world cases can be problematic. Often the most practical option to obtain a value of the integral would be a numerical method to begin with. This can take various forms, with some examples here. All, however, have in common that they right away create and then compute an approximation.

For various purposes, modeling aims for the total numerical value of the change experienced in a process. Formally, this would call for computing *definite* integrals (see Chapter 1: Change/differential equations), but in much of real-world modeling, numerical integration is often the realistic option. To see how it might work it is useful to turn, once again, to the geometrical aspect. As understood in calculus, a definite integral of a function equals the area between the curve and the *x*-axis between the limits of integration. One could, as is common in calculus textbooks, create an approximate counterpart of, say, a somewhat irregular curve using horizontal segments and obtain a value from the sum of the underlying areas. This can be improved upon by placing consecutive points on the curve and connecting them by straight lines, each with a slope of its own. The approximation would replace the rectangles with trapezoids as seen in Fig. 3.4 and allow to compute integrals numerically employing the *trapezoidal rule*. The area of each segment is taken to equal the distance between adjacent points x_i and x_{i+1}, symbol, h, times the *average* of the corresponding values of y, namely $(y_i+y_{i+1})/2$, as given by the formula $h_i(f(x_i)+f(x_{i+1}))/2$. The value of the entire integral, I, combining segments with equal distance, h, can be given by Eq. (3.9):

$$I \approx h\left(1/2\,f_1 + f_2 + \cdots + f_n + 1/2\,f_{n+1}\right) - 1/12\,h^3 \sum_1^n f''(x_i) \qquad (3.9)$$

where the coefficient ½ inside the parenthesis is appropriate for the end points, the first and last values, f_1 and f_{n+1}. To provide an example of an error/remainder term, generally disregarded here, the formula also includes the term on the far RHS.

Methods described so far essentially begin as polynomial linear interpolation procedures extending on Eq. (3.6), but then instead of calculating the value of a point on a curve, they evaluate the associated area as defined in integration. In actual practice, interpolation points are regularly placed not evenly, but in optimized patterns; to avoid issues associated with having many interpolation points, entire curves are computed piecewise. Not surprisingly, some sophisticated methods obtain polynomials for integration using

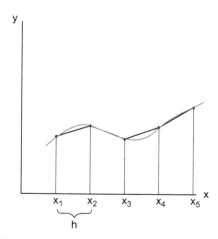

FIGURE 3.4 Trapezoid integral.

Lagrange interpolation. In that case the numerical method can be expressed concisely by Eq. (3.10):

$$\int_a^b P_n(x)dx = \sum_0^n \int_a^b L_nf(x_i) \qquad (3.10)$$

where symbols have meanings as above, Eqs. (3.7a,b,c). Substantially more advanced approaches are often required as documented in an extensive literature. While numerical differentiation and integration methods indeed can be instrumental in evaluating *ODE*s, cellular and other modeling commonly turn directly to the equations as such.

Values for differential equations

Numerical methods for evaluating *ODE*s have been a mainstay for some time in cellular and other real-world process modeling. Here to illustrate will be two basic, yet wide-scoped, prototypes for evaluating the ubiquitous, first-order, initial-value models. Related to this, both sidestep a formal solution process altogether. The first is an ingenious approach tied to geometry—historic, but still in use. The second offers popular approximation formulas in an analytical, computer-friendly format. Both approaches apply also to nonlinear *ODE*s and *PDE*s as well as systems of *DE*s. Both are represented in standard kinetic modeling software packages.

Tangents into curves/Euler methods

Euler's polygonal/tangent line methods were introduced in Chapter 1, Change/differential equations, in general terms. How they might work is emphasized here. Recall, first, that the solution values are those of points on curves approximating the "true" formal solution of the *ODE*. Constructing the approximating curves would be based on already familiar, simple ideas. (1) The procedure uses the initial values, y_0 and x_0, as its points of departure. (2) The slope of a curve at a particular point is approximated by the tangent. (3) The slope of the tangent is taken to equal its analytical counterpart, the derivative of a function. (4) And, here the crux, the notion that the local value, y, of an *ODE* at the same time represents the derivative of the solution and, therefore, the tangent and slope of the approximating curve.

To see how these ideas work in a most basic version, consider computing the solution of a linear, generic model, *ODE* equation $y' = f(x, y)$ given the initial value of both variables at x_0 and y_0. Using the initial values, x_0 and y_0, as the points of departure, the next points of the curve will be computed at consecutive values of x that increase by a constant value, h. That is at x_1 ($x_0 + h$), x_2 ($x_0 + 2h$), and so forth. Using a *forward* version of the procedure (see also Eq. 3.8a), the next values of the solution, y_1 at x_1, and those following, y_n at x_n, will be computed multiplying h by the *current* value of y as the slope, counterpart of the derivative of the solution. Starting at y_0 and computing y_1, the slope is simply taken to equal y_0, and y_1 thus equals $y_0 + y_0 h$. Similarly, $y_2 = y_1 + y_1 h$, and so forth. While compared to some of the previous alternatives in this chapter Euler's methods may seem crude, they work reasonably well with certain functions provided h, the x-axis segment, is sufficiently small. In more general practice, the procedure is often improved by further processing with another numerical method of which several are available. In that case, the first method is known as a *predictor* and the second as a *corrector*.

Taken together, the Euler procedure above can be summarized in equation form. The first step and each subsequent step separately obey Eq. (3.11a) below. A more general recursion formula that describes the entire stepwise procedure and is suitable for a computer algorithm is given by Eq. (3.11b):

$$y_1 = y_0 + f(x_0, y_0)h \tag{3.11a}$$

$$y_{n+1} = y_n + f(x_n, y_n)h \quad y_0 = c \tag{3.11b}$$

$$y_{n+1} = y_n + f(x_{n+1}, y_{n+1})h \quad y_0 = c \tag{3.11c}$$

where symbols have meanings as above, the subscripts n and $n + 1$ refer to consecutive points of the approximating function, and c is a constant.

Forward Euler methods such as that given by Eq. (3.11b) depend on the knowledge only of y_n and are also known as *self-starting* or *one-step*, reflecting the way they operate. They also have, and are known by, the mathematical property of being *explicit*, that is the independent and dependent variables are separate on both sides of the equation. However, in a common variant, the computation is conducted using Eq. (3.11c), replacing y_n and x_n on the right hand side of Eq. (3.11b) with y_{n+1}, x_{n+1}. This makes for a so-called *backward*, *nonself*, or *two-step* Euler method. It is also *implicit*, that is, the variables cannot be separated having y_{n+1} on both sides of Eq. (3.11c). Compared to forward Eq. (3.11b), the backward version depends on both y_n and y_{n+1} and hence is two-step and not self-starting. What matters here is that unlike y_n which is known, y_{n+1} is not yet known, and is calculated independently by a different method. Backward Euler methods can be more accurate and better with stiff *ODEs*, but evidently at the price of more computation. Both types are used in computer-modeling software packages.

Algebraic fine tuning/Runge–Kutta

Set originally to improve upon Euler methods, *Runge–Kutta (RK)* routines are popular. Algebraic and naturally computer-friendly, the *RK* approach has numerous versions for solving single and systems of *ODEs*.

Especially prominent is the (fourth order) *RK4* algorithm. Starting with a simple initial value *ODE* where $y' = f(x, y)$, $y(x_0) = y_0$, the *RK4* recursive formula is:

$$y_{n+1} = y_n + (k_1 + 2k_2 + 2k_3 + k_4)/6. \tag{3.12}$$

The sequence of k's on the RHS of Eq. (3.12) stands for: $k_1 = f(x_n, y_n)$; $k_2 = f(x_n + \frac{1}{2}h, y_n + \frac{1}{2}hk_1)$; $k_3 = f(x_n + \frac{1}{2}h, y_n + \frac{1}{2}hk_2)$; $k_4 = f(x_n + h, y_n + hk_3)$ where h has meaning as above and the ks are computed consecutively. The f here stands, as usual, for "function of" in the sense that if the original equation were $y = f(x^2, \sqrt{y})$, terms such as $x_n + \frac{1}{2}h$ or $y_n + \frac{1}{2}hk_2$ would be raised to the second power or taken the square root of, respectively. The formula also includes a not shown, separate error term on the RHS. As it were, the Eq. (3.12), *RK4* algorithm is a forward formula as were Eqs. (3.11a,b) of the Euler method. Both types are open to instabilities, but different versions may provide remedies. RK methods are mainstays of current *ODE* solving software.

Simulation

The term *simulation,* as has been pointed out in Chapter 1, Change/differential equations, can serve various purposes. Formally, models of the more complex cellular processes, having no analytical solution and being largely evaluated numerically, are considered simulations. Often simulation refers to a modeling practice in characterizing and designing real-world processes that are impossible or impractical to observe directly. These may be mimicked, instead employing conventional models or such with alternative methodology to explore and predict system behavior under conditions beyond those used originally to create the model. As will be seen in Chapter 8, Organized behavior/network descriptors, among others, there are also powerful simulation tools which replace or supplement traditional mathematical models with special apparatus of their own—Petri nets and Boolean models, for example.

Not confined to deterministic processes, simulation has also a role in modeling processes as random behavior using stochastic methodology. A prominent version, *molecular dynamics* simulation is a powerful, if resource intensive, methodology. It can, for example, describe at the amino acid or nucleotide level the motions involved in conformational change in proteins and nucleic acids or the atomic level motions of RNA polymerase along a nucleic acid, among others.

Simulations generally are carried out using computers. This offers not only enhanced capabilities to compute process values, but also ways to model special aspects of the time element. Common kinetic models describe processes that are active continuously, beginning and ending at the same time. Employing computers allows to initiate and terminate diverse processes independently. It allows to model also factors with delayed actions as well as altogether complex processes with staggered contributions, generally not practical with formal models. Computer simulations have also other benefits, for instance, allowing to incorporate conveniently real-time feedback about the status of a system.

Interfacing with machine computation

Quantitative modeling is intimately tied to machine computation as has been evident here and in previous chapters. Computers not only execute the instructions of a model; what they can compute, and how, has significant input into what these instructions will be to begin with. The choice of a model as well as its subsequent handling depend on the overall capacity of an available computer to perform the necessary operations. Memory required to store intermediate results may steer towards certain solution protocols and disfavor others. Speed and efficiency are top priorities in designing both hardware and software. Minimizing the cost of a computation is a routine concern. Realizing these objectives ties in with using simple, highly repeatable routines, computer algorithms based on recursion formulas such as those already encountered above and in Chapter 1, Change/differential equations.

Machine computation is generally prone to *error* of various kinds, obviously a major concern. Discrepancies between known or presumed "true" model values and the actual numbers generated can become significant when formal models, let alone approximations, interface with computer technology. With vast literature and associated terminology, here noteworthy is built-in, so-called *systemic* error. In the simplest terms, it is due to failure, one way or another, to include in the computation sufficient mathematical detail. Consider that kinetic process models employ continuous differential equations, but computers operate as machines that employ linear, algebraic, and *discontinuous* expressions. This eliminates decimals of a continuous value that depart from its discrete replacement and result in so-called *discretization* error. That some routines include only a limited number of a long sequence of terms can lead to *truncation* error. Underlying systemic error can be the technology, as such. The size of the physical template it uses, computer *word length*, imposes limits of its own, for example, on the number of digits, say, after the decimal point, that can be processed or excluding late digits causes *round off* error.

In addition to systemic error, the very act of computing can be problematic, resulting in *operational* error. Being physical systems, computers are susceptible to the *random* forces of nature, internal and environmental. There may also be shortcomings in human design and control of computer operations. What can be regular *noise* may be deceptive already, especially working with small numbers. Recall (Chapter 2: Consensus/linear algebra) that it can be unclear whether or not an *ODE* system has a meaningful solution, depending on whether a very small determinant of its matrix of coefficients is real or fictitious. Moreover, occasional error, notably in executing highly repeated routines, could propagate throughout and eventually become magnified intolerably, especially likely if it occurred early on. Of special concern in cellular process modeling is error associated with *instability* of so-called *stiff DEs*. Algorithms intended to generate good numbers cause operations to veer off course for no good reason with unreasonable results. Euler and RK methods, for example, call for computing a large number of intermediate points, each adding truncation error to an eventually unacceptable level. Operational error of all kinds can, thus, become another primary factor in choosing models, solution processes, and algorithms.

Keep in mind

Cellular models of ongoing process dynamics routinely encounter mathematical obstacles as well as demands of machine computation that call for *alternatives* to standard methods. Generally being *approximations*, these aim to provide numerical process values. In major versions, model continuous differential equations and their building blocks are substituted with algebraic equations, mathematically and computationally advantageous. With specific strengths and limitations, choosing an alternative method depends on the nature of the process, what already is known about it, and the purposes of the model.

Based on general mathematical principles, alternative series solution methods employ *polynomials*, sums of algebraic sequences. The *Taylor* series relies on derivatives and is effective mainly locally. Applications include linearizing nonlinear expressions, providing candidates for integration, converting refractory into computable expressions. *Fourier* series, consisting of sine and cosine functions and their integrals are natural choices for evaluating periodic phenomena, but are effective also elsewhere, generally in the large. A related *FT* can reveal periodic process components. *Chebyshev* polynomials approximate curves locally and globally, as well as provide for optimal design of other alternative methods.

Proved by experience, diverse *numerical methods* are regularly the practical or only option to solve for numerical values of *DEs*, their building blocks, and systems in cellular process modeling. Minimally formal, numerical methods generally derive from geometrical and analytical, readily computable model properties. Commonly computer implemented, they are routinely expressed as recursion formulas fit for a computer algorithm. *Interpolation*, *linear*, and *Lagrange* polynomials are appropriate when an approximation has to match precisely select points on a curve. The number and placement of interpolation points can be crucial. Numerical *derivatives* of functions may be obtained from the slopes of model curves. Values of *integrals*, *definite*, are obtainable from the area under a model curve using the *trapezoidal* rule. *ODE* initial-value problems can be approximated by the geometrical *Euler tangent* method, assuming that the slope of a model curve equals the derivative of the solution. Common in current modeling are the *RK* analytical formulas.

Using computers can affect model choice and handling. Issues are associated with operational capacity and design, notably memory size. Hardware and software aim for speedy, efficient, and low-cost operation. *Error*, a major consideration, can be *systemic*—generally a failure to accommodate sufficient mathematical detail. It can be *operational*, due to hardware and software shortcomings during a computation as well as the peculiarities of certain models.

Further reading

Dahlquist G, Bjoerck A. Numerical methods. Englewood Cliffs: Prentice Hall Inc; 1974. Mineola: Dover Publications; 2003.

Nikol'skiĭ SM. Approximations of functions. In: Aleksandrov AD, Kolmogorov AN, Lavrent'ev MA, editors. Mathematics: its content, methods, and meaning. Moscow: Russian Academy of Science Press; 1956. English: Cambridge: MIT Press; 1963; Mineola: Dover Publications; 1999. Chapter XII.

Pozrikidis C. Numericalal computation in science and engineering. New York: Oxford University Press; 1998.

Rivlin TJ. An introduction to the approximation of functions. Waltham: Blaisdell Publishing Company; 1969. Mineola: Dover Publications; 1981.

Tenenbaum M, Pollard H. Ordinary differential equations. New York: Harper and Row; 1963. Mineola: Dover Publications; 1985.

Chapter 4

Systems/input into output

Context

A major view of natural and artificial processes centers on a *scheme* whereby *input* enters a *system*, the active machinery of a functional unit, to become *output*. Practical in quantitative modeling the scheme is wide-ranging and already well-established. It is especially popular in technological applications where it has been mainly developed. Parts entering an assembly line turning into a car, instructions processed by a computer ending as an architectural blueprint, the people's will be legislated into law. A similar pattern can be seen in cellular processes at all levels, for example, amino acids assembled by the translational machinery into protein, hormone triggering signal transduction systems to stimulate cellular metabolism and differentiation, infecting virus recruiting cellular processes to produce progeny, antigen mobilizing cellular teams to elicit antibody formation, among others. As the technological and the biological examples show, the scheme can apply to the flow of materials as well as information.

Turning to the input/system/output scheme here will serve a number of purposes. Models in previous chapters generally and intentionally focused on the system component and its workings. Such follow the prevailing biochemical kinetics tradition, also when modelers with an engineering bent have the entire scheme in mind. Previous models, as already pointed out, can be demanding—with respect to data and resource requirements as well as the mathematics. They also may not provide all the desired answers. Adding perspective, perhaps altogether changing the point of view might be helpful. Moreover, the needs of cellular modeling go substantially beyond the survey so far. Describing models of the input/system/output scheme in this chapter will extend naturally on previous modeling infrastructure. It will be a first step in expanding its scope, which will be elaborated on in later chapters.

That biological systems and process behaviors, unique as they may be, can have analogs in behaviors elsewhere in nature and artificial world has been recognized already in antiquity (Icarus). There is thus a long-standing, still ongoing, dialogue between biological and technological disciplines. Here noteworthy is that processes in both environments commonly have input and output, are mediated by functional units that likely interact to various degrees, and that (deterministic) laws of change are such as those

Collective Behavior in Systems Biology. DOI: https://doi.org/10.1016/B978-0-12-817128-8.00004-3

delineated in Chapter 1, Change/differential equations. There are often shared organizational characteristics; among the more prominent are employing *modules*, often heterogeneous larger functional units dedicated to a particular task. Parallels are evident also in *design* and how devices work, with examples of technological control devices to follow in the next and later chapters. Not surprisingly, notions of technological origin such as those of system itself, feedback, and network, among others, also found their way into (cell) biological usage. Still other similarities, described by mathematical graphs and logical devices, will become apparent discussing networks and their processes in Chapter 8, Organized behavior/network descriptors.

The technological experience can be instructive also where the two environments differ. Cellular systems are prone to *change* during their lifetime, while the makeup, especially of engineering systems, more or less, will stay *constant*. Cells down to the lower subcellular levels are relatively *complex* and processes are highly *integrated*. Engineering systems are often simpler and easier to decompose and characterize as smaller subsystems. Moreover, *information* on cellular systems and their workings is still limited, but naturally available when technological systems are designed by humans. Altogether, technological undertakings have been historically more open to mathematical and abstract methods. Prompted by obvious mundane incentives, this made available an extensive body of instructive theory.

With roots in antiquity and forerunners in the industrial revolution the current wave of quantitative and abstract approaches to technological system behavior emerged in the mid-20th century. Under the umbrella of *cybernetics* these were initially combined with other responses to modern technological needs such as the theories of information, communications, and control. A more recent *general systems theory*, originating largely in biology and now broad-scoped, aims at an even more comprehensive view of the real world, but will have little application here. Ditto the systems methods of social and other sciences and related applications.

Directly deriving from technological practice, much more specific but still wide-ranging, is *linear systems theory* to be taken up here. Employing quantitative models based on the input → system → output scheme has a long history in engineering. It is also being applied already to higher-level biological systems, for example, in ecology and biotechnology process management, and is gradually finding direct applications in modeling cellular processes. Here valuable is also that linear systems theory offers a convenient opening to basic concepts in system approaches and to their wider context. The mathematics of the theory makes use of single and systems of equations of change together with alternative methods and linear algebra of the kind covered in previous chapters.

Turning to quantitative models, these will center on the two major approaches of linear systems theory—two distinct concepts in modeling. One, and increasingly in use, is the information-intensive *state variable (SV)* model which employs all three components of the above scheme. Being explicit about input and timing, it indeed extends the range of previous descriptions in this book to include outside interventions from the beginning of a process or when it is on its way, say, adding precursors or input from other processes.

Here especially interesting as a concept is the second approach. Offhand it would seem questionable to omit from a process description its natural centerpiece—its actual workings, represented by the system component. That, however, is what the *input/output (IO)* model essentially does. It turns out that modeling with only input and output can be valuable in its own right. Historically the earlier, the *IO* model, can still be effective when the workings of the system component are completely unknown or insufficiently known, situations not unfamiliar with living cells. More compact, the *IO* model may also be more convenient when the system component is of no immediate interest, as is the case in many applications. Similar situations, incidentally, underlie certain cellular applications (see complex whole/no parts, Chapter 1: Change/differential equations).

Control is a major factor in biological as well as technological processes. Essentially representing information that will manage material change, control is a main theme of the linear systems approach from its inception. Important as it may be, it is in effect viewed as a component of an otherwise already working system, a subprocess that will be dealt with in similar terms as the overall material process. The latter being the main focus in this chapter, will also prepare the way for dealing with the informational processes of control in Chapter 5, Managing processes/control.

Groundwork/descriptive/from input to output

What is at stake

A gist of the input/system/output scheme can be provided by turning first to quantitative and related questions that it may raise. Many of these relate to process dynamics as described in Chapter 1, Change/differential equations, and will not be repeated here. At the foreground is obviously how input and output relate, be it in the numbers of units that go in and come out, or in the respective economic values. Also of interest is how input and output depend on system and process characteristics—those intended originally, or perhaps as modified, say, by real-world conditions, by further changing the design, or by applying control. Turning to specifics—is there one input with one output, or are there multiples of either or both and in what combinations? Does a process depend on a one-time, initial supply of precursors? Will the manner of applying input, say instantaneous or distributed over time in various ways, be of any consequence? Would a system's previous history affect its response to new input? Will there be later input, possibly ongoing through the entire process lifetime? In more complex systems it may be pertinent what in an IO system can actually be monitored, estimated, and how. Could information on input and output only provide a handle on the way an insufficiently known system component itself functions? Questions to ask before choosing a model will be familiar from previous chapters. Staying deterministic, is change continuous or discontinuous (discrete)? Is it linear, its participating factors additive, at least over a workable range? Is the system homogeneous or heterogeneous? Do governing laws stay constant with time? How would they change, if not? Out of these and many other questions that can arise dealing with systems, a select few will be touched upon here from a technological perspective.

A word of caution before proceeding. The language, terminology, symbols, and overall style of linear systems and control theories originating in the technological experience are generally engineering oriented. Their usage sometimes differs significantly from those common in biological and related natural sciences. To maintain continuity and minimize ambiguity, with a few exceptions, this book generally follows the natural sciences conventions.

The cast

To serve their various purposes, literature definitions of the players in the systems scheme—input, system, and output—are flexible, highly context-dependent, and occasionally overlapping. *Input* and *output* will have their everyday meaning described below, but *system* needs more of an introduction. By some, a system is defined very loosely as that part of the real world that attracts someone's interest. Others have a system correspond to no more than a list of its members. Closer to the purposes of this book, as in Chapter 1, Change/differential equations, the notion of system will here refer to communities of cellular or other real-world functional units, homogeneous or heterogeneous, that perform a particular task collectively, that is, whether or not they interact directly within the system and possibly with the surroundings. This is in line also with the technological synonym for system, namely *plant*. Boundaries found in some traditional definitions of a system are not significant in this book.

Notions of input, system, and output naturally tailored to cellular and technological processes that convert precursors into products, such as those above, can apply also in other situations, among others when systems undergo change as such and exchanging materials with the environment is of no concern. An example is when one is modeling the diffusion or transport of a substance from one position (input) to another (output) within a cell. Or, for that matter, the classical case of a clock mechanism (system) that moves from midday (input) to midnight (output). Continuing with the process, the current position, formerly output, now will be input for the next step, and so on. Parallels in metabolic and signaling systems should be obvious. Input and output can be useful also as modeling devices needed to link together processes in different systems, each with its own description.

The play/a process on its way

What members of the cast actually do depends on the particular script. Scripts vary, but processes here will be taken to share two general features of their global behavior, their history over the entire duration.

First, in analogy with simple machine behavior, processes will be taken to be moving collectively forward in *one* overall *direction* all along. If one were, for example, to model glycolysis, that could mean disregarding branches that create amino acids or that secrete products from the cell. Such will be taken up discussing networks in Chapter 8, Organized behavior/network descriptors.

A second common feature is the observation that over time, real-world processes in progress typically are found in either of three *phases*. Often there is first a transition phase in which a system changes in value, say, a microbial culture in log phase increasing in weight or the intracellular amplification of external signals. It is known as *transient (phase)* in the technological context where it is backed by extensive theory often motivated by associated practical problems. A *transient* most commonly will lead to a constant-value *steady* or *stationary state*. The steady state is presumably the routine state of major cellular operations, as well as industrial production lines. Its properties are major factors in biological and technological success. Considered by theory to be a state-of-rest or *equilibrium* (chemical or thermodynamic equilibrium being a special case) the steady state also has an important role in modeling. Recall that it was key to evaluating parameters that govern system dynamics and stability behaviors, as seen in Chapter 1, Change/differential equations. Also noteworthy is that theoretically linear processes may have only one steady state, but nonlinear systems could have several. In the latter case, and depending on system parameters, equilibrium points may differ in properties such as stability and whether and where from they are accessible. That, in due course, processes may slow down to a halt can be less glamorous in real life as well as for theory. *Demise*, say, due to cellular components becoming dysfunctional, accumulation of toxic products, hostile invaders, or outright apoptosis is often represented as an active process in its own right with a transient phase and steady state of its own.

Many processes such as biological life cycles, cellular or higher, in fact do go through all three phases. This is not mandatory—various cell cultures, microbial or immortalized eukaryotic could conceivably be kept at steady state indefinitely. Except for the very first instant, an industrial production line could similarly start immediately at a constant pace, its steady state.

Periodic behavior, controlled by cellular level machinery is the hallmark of adaptive properties such as circadian and other biological cycles, as readers well know. There is also higher frequency oscillatory behavior that may have more specific roles in cellular life and other behavior whose biological significance is less clear, yeast glycolytic cycles being a classical example. Obviously oscillatory behavior is typical in nature and in human applications that are based on the wave properties of electricity and light. It is also a common hazard in technological systems with deficient control.

More time related

In addition to its obvious role in process dynamics, represented in models by the independent variable, there may be other time-related factors at work. For one, change that occurs in the system itself or its environment, say, due to activity modifying mutations as mentioned in Chapter 1, Change/differential equations, or perhaps creeping malfunction in a technological process. Technological modeling and increasingly that of cellular processes occasionally also deal with other time-related features, with two here being noteworthy.

Systems with a past

Although commonly avoidable in deterministic modeling, sometimes it is necessary to account for processes that took place before the period of interest. So-called bistable processes, for example, unlike more common processes encountered so far, can proceed along two different routes. Known also as *hysteresis*, quantitative behavior depends on previous history (i.e., which route was taken before starting the current activity). Bistable processes are used, for example, as *switches* in technological setups. A similar function is increasingly recognized in cellular genetic control; an extensively documented case is the mitogenic control system in the clawed frog *Xenopus laevis'* oocyte extracts. Technological systems with history-dependent output are thought to exhibit *memory*.

Delay

Response to input into real-life systems is often delayed significantly on the timescale of an ongoing process. For modeling purposes there is the formal option to turn to *delay differential equations*, in most basic format $dy/dt = f(x, t-\tau)$ where τ is the delay. Its simple appearance, however, is deceptive and delay equations can be difficult to solve. Delay in modeling is often managed using computer protocols.

A quantitative view

Two approaches to modeling processes predominate in linear systems theory. The SV and the IO models differ in substance as well as style.

Both, nevertheless, draw on the mathematics introduced in Chapter 1, Change/differential equations. Choosing between the two models depends on the purposes of the analysis and on system information that is available. Both the *SV* and *IO* models play a significant role when it comes to technological control, as will be seen in Chapter 5, Managing processes/control. The focus will be on modeling continuous processes while keeping in mind that discontinuous counterparts are common and generally follow similar logic.

State variable model/the cast in full

Given appropriate information on the system component, the *SV* model employs the entire cast (however, see below). The more comprehensive of the two models, SV is becoming increasingly popular in dealing with technological processes, material and informational (control).

System

Central to the *SV* model is that processes are seen as a succession of stages of an evolving system. Each stage corresponds to the condition of the system at a particular moment as specified by quantitative parameters thatformally make for a *state*. Analogous descriptions of matter in terms of pressure, temperature, and volume, parameters that can vary, *SVs* will be familiar. The more common state parameter in cellular process modeling is the current concentration of a particular component. The actual unfolding of a process is seen as an *updating* of its states. Describing processes as a succession of states is common modeling practice elsewhere—later on in this book, for example, in dealing with various network and Markov processes.

Technically this line of thought readily translates into describing deterministic processes once again by differential equations and their solutions— in this format the state, given by the current system values, is represented by the dependent *SV(s)*, x. Retaining here technological symbol usage, x replaces the familiar y used so far because in the systems theory literature the symbol y is usually reserved for output. The time, t, will as usual serve as the independent variable. In line with the mathematics the *SV* model also mandates and carries the advantages of providing initial conditions.

Input

The input/system/output scheme beginning this chapter stands for precursor input that is converted by a system into product output. The system component here uses what is known as *primary* input. Obviously, cellular processes, say glucose converted by a plastid into starch, or others described by Chapter 1, Change/differential equations, models, operate the same way. Taking a closer look at their models, for example Chapter 1, Change/differential equations, Eq. (1.1a) representing most others, the state of the process was described by the concentration of the product, output protein; input, say of amino acids, was assumed to be adequate but was not represented explicitly. The *SV* model is in the same mold.

In addition to carrying out a primary input-dependent ongoing process, technological systems often employ *secondary* input that can affect the state and eventually the output of the system. The *SV* model thus extends on Chapter 1, Change/differential equations, models with an extra provision for representing secondary input in the form of another variable, u, here being a generic symbol for both types of input. It most notably serves to represent control as discussed in Chapter 5, Managing processes/control. In fact, reflecting that usage, secondary input is also referred to in the literature as *forcing* input making for a *forced* system. Absent forcing input the system is *free* running. Such, by design, were for example almost all the kinetic models shown in Chapter 1, Change/differential equations. Note, incidentally, that using the state and time as variables, the model offers a way to specify secondary process timings.

Output

Variable y refers here, again, primarily to the material product of a process. Seen more broadly, output corresponds to a reading of the current status of a system. That could be direct, for example, of the value of a concentration or a transmembrane potential, or indirect, say when assayed as the toxicity or antibiotic activity of a product.

At work

The *SV* format in line with Chapter 1, Change/differential equations, modeling is geared to processes that again are additive (i.e., linear or linearized) as well as such that maintain a constant behavior during a process and are known as time invariant (*autonomous*) or others whose behavior changes and thus are time-dependent. With x, u, and y as dependent variables and time, t, as the independent variable, the actual *SV* model employs two types of statement used together or independently, as the case may be. State updating, linking current system status and input will be given by single or systems of differential *state equations*. Output, will be related primarily to the updated state using algebraic, again scalar or matrix *output equations*.

State updating

As in Chapter 1, Change/differential equations, modeling favors linear, time independent, first-order, ordinary differential equations (*ODEs*). Employing above terminology, the state equations consist of a term corresponding to a "free" contribution, x_{free}, and a "forced" contribution, u_{forced}. In the simplest, general case there is a single process with one each of a primary input, a SV, a secondary input and an output. Consider generating a substance partially deactivated by an inhibitor:

$$dx(t)/dt = x_{\text{free}} + u_{\text{forced}} = ax(t) + bu(t) \quad x(t_0) = x_0 \qquad (4.1)$$

With state and input variables x and u, the coefficients a and b provide for values specific to the process of interest. Here constants, the latter may be made time-dependent, that is, ($a(t)$ and $b(t)$) if needed. Separately on the right hand side (RHS) of Eq. (4.1) are the initial conditions as before.

Readers will recognize that the second term on the RHS, the forcing term, corresponds to the *nonhomogeneous* term, $Q(x)$, of Chapter 1, Change/differential equations, Eq. (1.3) ODE prototype. In real life forcing input may be of two types. One is a controllable component, u. The other is uncontrolled, often random, due to internal causes or the external environment, and denoted separately, say, by w. It may be advantageous to keep the two separate during further manipulations.

A system with multiple simultaneous subprocesses, represented by multiple variables, is *multivariate*. Also encountered in Chapter 1, Change/differential equations, such commonly call for a matrix version of the state equation and further management by the methods of linear algebra. The vector/matrix, boldface counterpart of Eq. (4.1) then is:

$$d\mathbf{x}(t)/dt = \mathbf{A}\mathbf{x}(t) + \mathbf{B}\mathbf{u}(t) \quad \mathbf{x}(t_0) = \mathbf{x_0} \qquad (4.2)$$

where $\mathbf{x}$ and $\mathbf{u}$ are the vectors that group but keep track of individual state and input variables, and $\mathbf{A}$ and $\mathbf{B}$ are the respective matrices of the coefficients. Note that the matrix version of the state equation also provides initial conditions independently for each SV grouped as a vector.

Output

Output in the *SV* model is given by a straightforward, algebraic, *output*, or *readout equation*. There are again, depending on the case, univariate or multivariate versions, as shown next:

$$y(t) = cx(t) + du(t) \qquad (4.3)$$

$$\mathbf{y}(t) = \mathbf{C}\mathbf{x}(t) + \mathbf{D}\mathbf{u}(t) \qquad (4.4)$$

where y and $\mathbf{y}$ represent single and multiple outputs, other variable symbols have the same meaning as above, and $\mathbf{C}$ and $\mathbf{D}$ are again coefficient matrices. The second terms on the RHSs of Eqs. (4.3) and (4.4), $du(t)$ and $\mathbf{D}\mathbf{u}(t)$, have their uses in modeling certain processes and their control, but are of no further interest in this book and can be safely omitted.

Computing the output, as can be seen in Eqs. (4.3) and (4.4), first on the RHS, requires the terms $x(t)$ and $\mathbf{x}(t)$. These are as yet not available as such, but have to be obtained by first *solving* state Eqs. (4.1) or (4.2). These, unlike major Chapter 1, Change/differential equations, kinetic models that use homogeneous *ODE*s, now include a "forced" contribution, u_{forced}, and are nonhomogeneous. Such can be solved formally, for example using Laplace transform (see Chapter 1: Change/differential equations). The actual solution of the most elementary state Eq. (4.1) is given by the equation on the far RHS of Eq. (4.5):

$$x(t) = x_{\text{free}} + u_{\text{forced}} = e^{at}x_0 + \int_0^t e^{a(t-\tau)}bu(\tau)\,d\tau, \qquad (4.5)$$

where τ is the dummy variable of integration, a technicality of no concern here, the variable that counts being the time, t, as shown with the integral symbol. In effect the solution combines the solution of a homogeneous *ODE* given by $e^{at}x_0$ and familiar from Chapter 1, Change/differential equations, and the somewhat formidable contribution of the nonhomogeneous term, the integral next.

To obtain the output, $y(t)$, will take substituting the expression on the RHS of Eq. (4.5) into the output Eq. (4.3), followed by moving the expression e^{at}, common to both terms, before the brackets. This will yield:

$$y(t) = ce^{at}\left[x_0 + \int_0^t e^{-a\tau}bu(\tau)d\tau\right] + du(t). \qquad (4.6)$$

There are similar solutions to matrix versions (not shown) where the constant a is replaced by the matrix A.

The practical goal of putting a model through a solution process, as will be recalled, is often to obtain actual numbers or *numerical values* for state descriptions and output. This calls for computing the value of the exponential expression e^a by inserting actual values of the constant a in both terms and is straightforward. It is, however, potentially problematic with matrix versions that represent multiple inputs and outputs because evaluating the expression, e^A, a *matrix exponential*, can be difficult. Once again, computer-implemented, numerical, solution methods of the type described in Chapter 3, Alternative infrastructure/series, numerical methods, can be handy.

Input/output model/success with less

How process output would respond to input is obviously a primary consideration in evaluating process performance. In various circumstances—biological, technological, and economic, among others—information on the workings of a process is lacking, partially or completely. Yet, returning to the original input/system/output scheme, that which uses primary input, modeling with input and output only can be still useful and is regular practice. Employing the *IO* model the system component is thus taken to be a *black box* and is altogether disregarded. Taking this route can be a necessity, but can also be a matter of choice. For many practical purposes—bioassays or various technological setups—how output relates to input may be all one needs or wants to know. Less detailed than the SV model, the *IO* model is considerably more compact and may be advantageous especially dealing with complex interactions.

Having originated largely in electrical engineering and earlier historically than the *SV* model, the *IO* model is still popular in technological modeling. It is also finding its way into cellular modeling with recent applications, for example, in studies of electrical properties of nerve cells and of signaling systems elsewhere. The tools described here will also have a substantial role in control modeling infrastructure to be taken up in Chapter 5, Managing processes/control.

Quantitative

The *I/O* model is geared toward linear processes whose behavior does not change throughout their lifetime. Focusing once again on continuous processes as the prototype, input and output are described initially by autonomous, linear, differential equations assuming initial values equal zero. Unlike *SV* model equations, used as is, *IO* model equations are subjected first to the *Laplace transform* (see Chapter 1: Change/differential equations). Advantageous for mathematical work up, recall that this puts *IO* model equations into the algebraic *Laplace*, or $s-domain$. Instead of depending on the time, t, terms now formally depend on the variable s, a technicality that needs to be acknowledged because it appears on the page, but is here of no consequence otherwise.

At the center of the *I/O* approach is the direct relationship between output and input. The relationship is expressed by a *transfer function* defined as the *ratio* of output to input. Given a system with a single input, $y(s)$, and a single output, $u(s)$, the transfer function, $h(s)$, is $y(s)/u(s)$ (Eq. 4.7a) with (s) again denoting the Laplace domain. For multivariate processes, with multiple inputs and outputs, there are counterparts in the *matrix transfer function*, $\boldsymbol{H}$ (s), Eq. (4.7b). An output formula, available in both versions but shown in matrix form only, is then given by Eq. (4.7c), a rearrangement of Eq. (4.7b). Thus,

$$h(s) = y(s)/u(s) \tag{4.7a}$$

$$\boldsymbol{H}(s) = \boldsymbol{y}(s)/\boldsymbol{u}(s) \tag{4.7b}$$

$$\boldsymbol{y}(s) = \boldsymbol{H}(s)\boldsymbol{u}(s) \tag{4.7c}$$

Going through the motions of the solution process in the Laplace domain, the final outcome, nevertheless, will be obtained in a time dependent format by applying the *inverse* Laplace transform (see Chapter 1: Change/differential equations).

In addition to IO formulas, *IO* models can also be a source of other process information. Once in the Laplace domain *IO* equations will generally be in the form of a *ratio* of two multiterm expressions, polynomials (see Chapter 3: Alternative infrastructure/series, numerical methods). Taking it one step further, the *numerator* and the *denominator* may be considered as two independent equations with *roots* computable when each is independently set to equal zero. The roots of the denominator, also known as the *characteristic* equation, are here known as the *poles*; in the multivariate case they are none other than the system eigenvalues (see Chapter 1: Change/differential equations). Transfer function poles are, thus, intimately related to system dynamics and stability behavior and in fact serve as important benchmarks in implementing control. Roots of the numerator equation, the so-called *zeros*, may be indicative of transient dynamic behavior.

Note also that: (1) Transfer function symbols other than $h(s)$ and $H(s)$ are common in the literature. (2) Although input and output are represented by a single symbol in Eq. (4.7a,b,c), y and u in real-life models can stand for multiple-term equations to start with as was the case in Chapter 1, Change/differential equations. (3) By the conventions of linear algebra (see Chapter 2: Consensus/linear algebra) $H(s)$ in Eq. (4.7c) is always a premultiplier (on the left) of $u(s)$.

Between models

Altogether, *SV* models have the benefit of a wider range of quantitative methods while the *IO* model has the advantage of being simpler and more compact. Could working with an *SV* model benefit also from the advantages of the *IO* model and vice versa? Both the *SV* and the *IO* descriptions in fact originate as single or matrix differential equations. Could they be interconverted? That the more information rich *SV* equations may serve as origin for *IO* transfer functions with less information, offhand, would seem feasible. It can be done by substituting *SV* expressions for output and input in a transfer function. Developing *SV* counterparts to *IO* models is indeed more of a challenge. However, detailed analysis shows that in favorable cases transfer functions still retain sufficient information on input, system, and output to make this possible with the help of certain additional assumptions.

Graphical aids

Mobilizing the visual experience by using graphical devices is a long-standing practice in modeling, especially of complex systems and processes. A major topic in its own right, it will be taken up in some detail in Chapter 8, Organized behavior/network descriptors. As will be seen there, graphical devices can provide overview as well as tools for dealing with detail. Two types of diagrams are prominent in linear systems and control modeling where they serve to represent complex systems visually as well as to facilitate actual computations. Both can be used with the *SV* and the *IO* models, the bigger and the lesser. Depending on the purposes of the modeler both types of diagrams can be employed separately or together.

Block diagrams

Probably familiar, block diagrams such as Fig. 4.1 are often mobilized to represent higher-level activities of complex functional units. They are popular for instance in modeling technological control processes, as will be seen in Chapter 5, Managing processes/control. As shown in Fig. 4.1, these employ two main types of building block, or *element*. *Rectangles* represent the active unit and *circles* represent *summation points* standing for one-time addition or subtraction of inputs. When needed there are also *pick off* or *branch off* points, perhaps represented by a smaller circle, a dot, or no object at all. Fig. 4.1 could for example represent a process with two inputs and two outputs. Input u_1 would represent a precursor generated previously and u_2 could represent input from another reaction, perhaps (anticipating Chapter 5: Managing processes/control) a control, activating or deactivating. These inputs are combined at the summation point (circle) to determine the output of the system (rectangle). In this case, the overall output, y, is at the pickoff point (dot) split in two, y_1, and y_2, with a different destination each as there are two different users. Certain block diagrams may include special elements, for example to represent *delay* (not shown). Connecting the building blocks are lines or arrows that link elements that interact and specify the direction of input and output at each step.

In quantitative modeling the rectangles usually carry an expression that represents the activity of the functional unit, for example a transfer function or a rate expression. In some cases it may be possible to simplify calculations by consolidating diagrams according to so-called *reduction* rules. Among others these specify how to combine formulas of units that act in parallel/concurrently or in series/consecutively as well as how to favorably relocate summation and pickoff points.

Signal-flow graphs

Modeling processes such as for example multisubstrate enzyme reactions or signal transduction cascades with their numerous interactions, there may be

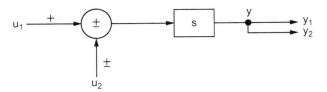

FIGURE 4.1 Block diagram.

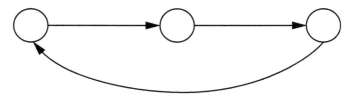

FIGURE 4.2 Signal-flow graph.

special interest in the *direction* of material or signal flow at each elementary step as well as in the overall configuration. Visual representation in addition to tools for further analysis are available in *graphs*, here a preview of what will be taken up in some detail in Chapter 8, Organized behavior/network descriptors. As graphs typically do, signal-flow graphs, Fig. 4.2, center on two elements. (1) *Nodes* or *vertices* that here stand for functional system members, say a pathway intermediate, and shown as little circles. And (2) *edges* or *branches*, lines with a direction (*arcs* in Chapter 8: Organized behavior/network descriptors), connecting the nodes to represent functional relationships between system members, say one being a source of input to the next pathway member. Again, anticipating Chapter 5, Managing processes/control, Fig. 4.2 depicts a system with three elements where the third provides feedback to the first. As with block diagrams, signal-flow graphs can be simplified using rules of their own.

Signal-flow graph elements can be matched with building blocks of mathematical equations. Nodes would correspond to variables and edges to the formulas that relate the variables. This opens up a range of applications in which system problems are initially represented by graphs, but are solved converting them into equations and vice versa. Elements of signal-flow graphs also can correspond to, and be interchanged with, those of block diagrams.

Especially notable here is *Mason's (gain) rule*, a formula not shown that is handy in modeling complex processes. Applied to throughput in pathways with multiple inputs and outputs, it greatly simplifies otherwise complex calculations. Based on particular features of a process' signal-flow graph, Mason's rule allows to select among many interactions possible in a system, only a few being critical to the main process. Having originated in electrical engineering and now in wider use, a classical application of Mason's formula to cellular systems is in the King—Altman approach to the rate behavior of enzymes with multiple partners (i.e., substrates, effectors).

Keep in mind

Cellular process modeling and its infrastructure benefit from extensive technological experience. A scheme whereby *input* is processed by a *system* of functional units to become *output* captures essential characteristics of process behavior at various levels in both environments. Based on this scheme, linear systems theory is a major quantitative approach to modeling in the real world, and especially technological, processes. It is applicable to processes that transform materials as well as transfer information and can deal with collective behavior of homogeneous or heterogeneous systems. Where applicable its elementary applications provide models for routine operations, various interventions, response to short and long-term external factors, and cost/benefit estimates, among others.

Deterministic, the main mathematical tools of continuous linear systems theory are *ODEs* and their systems as managed with linear algebra. Modeling options depend on information available. Given data on input, system evolution, and output, the more explicit *SV* format may be preferable. Its mainstay is the *state equation* describing processes as a progression of instantaneous states of a system. Solving this equation allows to separately compute an *output equation*. Using knowledge on input and output only, the *IO* model is a black box approach. At its center is the *transfer function*—the Laplace domain ratio of output to input. Different appearances notwithstanding, the *SV* and *IO* models are related and possibly interconvertible. Eigenvalues and poles of the respective *SV* and *IO* equation systems are related to system dynamic and stability behaviors.

Visual infrastructure for representing systems and processes and handling their models includes block diagrams and signal-flow graphs. Both graphic aids come with rules of their own and can be handy in facilitating calculations of system and process properties in *SV* and *IO* models, Mason's rule allows to calculate throughput of complex systems focusing on process essentials.

Further reading

Auslander DM, Takahashi Y, Rabins MJ. Introducing systems and control. New York: McGraw-Hill Book Company; 1974.

Chen C-T. Linear systems theory and design. 4th ed. Oxford: Oxford University Press; 2013.

Dorf RC, Bishop RH. Modern control systems. 11th ed. Upper Saddle River, NJ: Pearson Prentice Hall; 2008.

Gopal M. New Delhi: New Age International Publishers, Reprint Modern control system theory. 2nd ed. New Delhi: Wiley Eastern Limited; 2010.

Chapter 5

Managing processes/control

Context

Biological cells function normally when the contributing processes are on track, on time, coordinated, and dependable. Were cellular processes entirely on their own, meeting these or other more special conditions would be unlikely. That processes behave well collectively takes superimposing the powers of *control*. Facing similar issues, control is ubiquitous also elsewhere in nature and in the artificial world as readers well know. Light initiating a chemical reaction and a switch turning on a motor; global temperature governing long-range ocean levels and a central bank regulating the national money supply; death ending life, a rocket exploding itself out of existence. Process control in biological cells serves similar objectives (see Quantitative views in this chapter, below) and is evident at all organizational levels. Among the better explored are control of metabolic pathway activities by their own members, triggering and regulating the machinery of gene transcription by nutrients, guidance of the transitions of the cell cycle by special signaling assemblies.

Activating a biological function, a choice is made first *which* particular process(es) is/are to be set in motion, say, of one particular out of various energy providing pathways, or of entire systems to be moved from a growth to a reproductive mode, among many others. A matter of control as well, but a qualitative decision, it will here be taken for granted. In line with the main themes of this book the focus in this chapter turns to the ensuing quantitative and time aspects of control: how *much* and *when* a process will be active, *how* control can bring this about. It will also be an opportunity to see how the linear system modeling in Chapter 4, Input into output/systems, can be used, extend on its infrastructure, and introduce an altogether new methodology.

A basic level of control is already evident in the collective behavior of cellular processes that are *self*-regulating. System *members*, say, of a metabolic or signaling pathway, control the activities of their nearest (and possibly other) pathway members and are, in turn, subject to similar control by other system members. Heavy car traffic in which each car can move only depending on the others might be an analog. Control in these cases is built into the process and is typically independent of external factors. Yet, control of major cellular as well technological processes is commonly mediated by

Collective Behavior in Systems Biology. DOI: https://doi.org/10.1016/B978-0-12-817128-8.00005-5

more sophisticated, higher, *special structures*. These are distinct from the working system and respond to special cues as seen in the other examples. While the two prototypes will be discussed here separately, actual control of cellular processes often combines both. In fact, as will be seen in a later chapter, cell cycle special control structures are self-regulating.

Turning to *model* the quantitative and time aspects of control in cellular processes—whether controlled from within or from elsewhere in the cell—customized kinetic models in Chapter 1, Change/differential equations, are considered the gold standard. In fact, characterizing control and its potential response to change in process makeup has always been the incentive for creating major models in the first place. However, already demanding technically and theoretically, to provide a full account of observed process behaviors, models may still require additional infrastructure. Responding to diverse needs, current modeling of cellular control is thus eclectic and open to innovation. Input from elsewhere could be useful.

There is indeed extensive technological experience with control. Over time this has generated general concepts as well as specific methods, as will be seen in this and later chapters. Here it will be instructive to observe first some of the features common to the biological and technological environments and where they differ, as in Chapter 4, Input into output/systems. In both cases control is used to ensure that processes follow their particular plan and perform as delineated at the beginning of this chapter. The cost of control—metabolic, genetic, or in a technological setting—is often significant and, thus, another shared factor. There are also analogies in the way control operates. Cells as well as technological systems employ controls that use real-time process information as feedback and its specialized apparatus and others that do not. Parallels are evident also in design detail and in computing process ongoing response to irregular activity levels, as will be seen. On the other hand, unlike biological cells with their highly integrated, predominantly chemical processes, technology frequently brings together largely independent and often separable forces. Steel production, for instance, combines practically independent thermal, mechanical, and chemical factors jointly guided by electromagnetic signals. Simpler to deal with and backed by information on design and the materials involved, technological processes are, offhand, relatively easier to analyze and experiment with, let alone present at an elementary level.

Major approaches to modeling technological processes build on the linear systems models presented in Chapter 4, Input into output/systems, and readers are assumed to be familiar with these. Wide-ranging applications extend from evaluating the overall feasibility and consequences of controlling an entire process, down to establishing technical detail. Among the more typical questions that arise are: Is a particular process indeed controllable? Do limits on control exist and what would they be? Would employment of special controls improve a process sufficiently to justify the extra cost? If so, where in the system should controls be placed? How much control would be required? How will it affect overall system behavior? Will there be adverse consequences? Is there an optimal way to allocate limited resources between system and controls? When it comes to cellular processes and their long Darwinian evolution, questions such as these are obviously after-the-fact. Yet the same factors, here in a technological context, most likely—in the past and continuing into the future—are at play also in cellular biology. However, these questions and potential applications in a cellular setting, still largely remain open.

Infrastructure for linear modeling of process control will be introduced in the first part of this chapter. Addressing select issues in control, this will expand on infrastructure in Chapter 4, Input into output/systems. Context will have here a significant role as it directly relates to the overall objectives of control, the types of information used, and how control is deployed—all of which are intimately tied to the configuration of each system. A common distinction here is between control that has been *preprogrammed* and is associated with an *open-loop* configuration and control that employs *ongoing* information as *feedback* in *closed-loop* arrangements. Both configurations are open to state variable (*SV*) and input−output (*IO*) modeling discussed in Chapter 8, Organized behavior/network descriptors. Also evident will be a role for organizational structure, the second main theme in this book, and its associated systems design.

Modeling technological process control, the linear approach is largely applied to control superimposed by special devices. It also informs cellular modeling, however, notably of self-regulating process control. *Metabolic control analysis* (*MCA*) is a unique approach geared toward control in common multienzyme pathways, but applicable also to other cellular systems. Its origins are in (see Chapter 1: Change/differential equations) kinetic models created bottom-up. Incorporating technological input allowed to create a new and innovative top-down modeling concept that allows to address new problems and reexamine old ones. Already covered extensively in the literature, a brief outline of *MCA* in the later part of this chapter will serve to introduce the special infrastructure it has generated as well as its technological connections. On the other hand, modeling special device−mediated cellular control, the mainstay of cellular process control, commonly turns in different directions. Conventional kinetic models for one, but prominently also to models that employ a spectrum of yet other infrastructure. In this book the discussion of this can be found mainly in Chapter 8, Organized behavior/network descriptors, covering networks, and Chapter 6, Best choices/optimization, on optimization methods.

Control structure and function

Control is a broadly understood concept. The *term* control could imply here an *activity* that physically modifies behavior in a system, or a *device* that carries this out. Much of the quantitative discussion here will concern control as an *effect* on process behavior. In that sense the term *regulation* in cell biology would refer to a *consequence* of control. However, before actually turning to the quantitative and time aspects of control it will be worthwhile first to take a look at what, in reality, it does and for what purpose.

Targets and immediate objectives

Consider first processes taking place without control, sometimes called *free running*. In many simple, technological processes functional components of a system convert input into output at fixed settings. Suppose that for some reason it is desired to affect output, perhaps to change the amount of product as planned or to correct for unintended interference. Offhand, adjusting the input could be an option. This, however, is generally a crude and inefficient type of control, be it cells or industrial processes. In fact, cellular and technological systems both generally manage the *machinery* (the *system* from Chapter 4: Input into output/systems) instead of the input using specially devised control. In other words, the output of the *primary* process that converts material input will be adjusted by control operating as a *secondary*, control process. Although still physically mediated, say chemically, electrically, or mechanically, control here serves as *informational* input into the system—an input that is *forcing* the primary process from its current behavior to the one designated, eventually affecting its output.

Technological control usually aims for a particular or combination of readily identifiable objectives. These have analogs in cellular control where they are likely to be more integrated and less neatly separable. Here four are noteworthy. (1) *Triggering* or *terminating* particular activities. In a cellular version (e.g., hormones, nutrients, or light triggering-gene expression) cellular cyclins activating cell cycle control machinery, or cytochrome *c* initiating apoptosis to terminate cellular life. (2) *Steering* system transitions from one state to another, for example, electric potential that moves a neuronal membrane from a polarized to a depolarized state or cellular control machinery directing cell cycle transitions. (3) *Maintaining* a predetermined *steady state* so that, facing tendencies to deviate, output is stable and deviating minimally from a target value—for example, feedback in self-regulating metabolic pathways or, repair, degradation, and replacement processes that maintain cellular makeup. (4) *Responding* to a change in environment, internal or external (e.g., the appearance of toxic substances) by activating cellular detoxification mechanisms or pathogens neutralized by innate cellular immunity.

Schematics

Numerous technological processes are satisfactorily free running as designed. When even minor intervention is needed, attaching extra controls is often not trivial. Extra expenses, questions of technical feasibility, and negative side effects may force partial or even complete redesign of an entire process. Shortcomings of a technological control mechanism itself in monitoring and correcting system performance may also require yet another layer of response.

With counterparts in biological cells, applying control to technological processes generally falls within two prototypes. One takes into account the starting conditions and aims for a preset target output. The other, in addition, relies substantially on status information obtained while the process is taking place. Both types of control come with custom devices and differ in their design, the feature by which they are usually known.

Open loop

In the first prototype and the older technological option, control is *pre*programmed. Less costly, but also less accurate, it is appropriate given sufficient certainty about the initial conditions and subsequent behavior. Open-loop control can be applied in different ways—in some versions only once, often at the beginning of a process. In others, still preprogrammed, it is applied intermittently or continuously while a process is unfolding. Open-loop control also has technical advantages that include simplicity and an immediate effect without time delay.

A most elementary setup of technological open-loop control is shown in Fig. 5.1A. Focusing here and in Fig. 5.1B on the control structure, the primary input, say of a material precursor, is not shown with the understanding that it feeds independently into the system component, the active functional unit. As shown, control in the Fig. 5.1 open-loop system originates with a device that provides *control input* based on the target, or *reference* output of the (primary) process and on the initial status of the system. This information is received by a *controller* that delivers a predetermined response to keep activity on track.

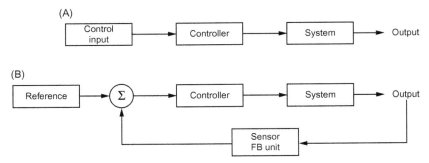

FIGURE 5.1   Open (A) and closed (B) control loops.

Open-loop control is common in cellular processes as well, as will be seen especially in Chapter 8, Organized behavior/network descriptors. Among the prominent are *feedforward* loops where—albeit involving components additional to those shown in Fig. 5.1—cues act at one end, control takes place at another. Among higher-level examples are nerve impulse−triggered cascades starting with the release of calcium into the cytosol anticipating muscle-cell contraction. At the lower-molecular level forward loops are encountered in enzyme catalysis and notably in prokaryotic and eukaryotic genetic control of metabolic and other pathways (see network motifs in Chapter 8: Organized behavior/network descriptors).

Closed loop

In the second and more recent technological option, real-time information is collected on the ongoing status of a process and used as *feedback* to evaluate whether the process is on track and make the necessary corrections if it is not. *Positive* feedback enhances process activity; *negative* feedback is deactivating. Turning to the closed-loop setup, actual control is delivered by counterparts to the open-loop control elements described above, though with additional arrangements. In the example shown in Fig. 5.1B these include a *sensor* or *feedback* device that measures the output and feeds it back to a *summation* device. (The summation symbol, $\sum$, is here an alternative to the $\pm$ sign in Fig. 4.1). The summation device is where the discrepancy is computed between the actual output and that intended as provided by a reference value. This so-called *error*, if any, is communicated to the controller for subsequent operation. As can be seen, together the components of the machinery indeed form a closed loop.

Closed-loop control is normally more effective than the open-loop configuration when facing unpredictable system behavior. It is, however, generally more expensive, complicated, and in the technological experience, more prone to undesirable side effects. Negative feedback can be *destabilizing*, a problem notably in delivering electric power. Especially prone could be processes with quantitative makeup that favors oscillatory behavior (see Chapter 1: Change/differential equations). Instability could also be due to practicalities such as time delays when data collection is slow or when the mechanics impede the transmission of a control's response. Positive feedback can be excessive, ending in hazardous loss of control, notably in handling explosive chemicals or in conducting preplanned forest fires.

As readers well know, feedback control is ubiquitous in cellular activities. A familiar classic and especially well-documented example is the *negative* feedback of lactose when present in excessive amounts on the activity of the *Escherichia coli* lac operon. *Positive* feedback is thought to play a major role, for example, in switching between alternative processes in so-called bistable systems, be it in glycolysis or in preserving memory of the past.

In Fig. 5.1A the different functions that together make for feedback control are nicely separated. The way feedback control operates in the more integrated cellular systems is, nevertheless, in many ways similar. This is not surprising recognizing that cellular processes are carried out by components that, among others, have properties of control elements. Reckon that process initial steps such as ligand binding to a receptor or a substrate to an enzyme in effect sense their concentration in the environment. Interactions at the end of a process such as the release of a ligand, an enzyme product, or a cellular signal are similar. Moreover, response is measured using a reference to determine the direction and rate of an interaction—possibly an equilibrium, perhaps a steady state. In other words, an enzyme molecule, for example, like a technological counterpart, could act as a sensor, transduce information as well as compute and activate a response. Products could provide information for coordinating activities of other system members or elsewhere in a cell. In fact, it is control properties such as these that provide for the ubiquitous self-regulation in cellular pathways.

Quantitative views

Models

Given that primary processes are linear, both open and closed-loop control may be evaluated using linear systems methods presented in Chapter 4, Input into output/systems. Either the *SV* or the *IO* model or a combination of the two could apply, depending on the case. Turning here once again to the more common continuous versions, keep in mind that discontinuous analogs have significant usage as well. Essentially following similar logic, these will not be discussed here separately.

Control over a single process (or another system parameter) is known as *single-variable* control. Control that targets multiple processes in a system makes for *multivariable control*. Matching equations of Chapter 4, Input into output/systems, will be useful here. This, however, with the understanding that input, symbol u, now is not material but stands for information that modulates the activity of the system and eventually the output. Single-variable control has a match in a single, scalar equation. Multiple-variable control will be represented by multiple, simultaneous equations, frequently in vector and matrix format. The latter allows to keep track of participating processes, assure the correct matching of processes with controls during the mathematical work up, as well as offer access to the methods of linear algebra (see Chapter 2: Consensus/linear algebra).

The state variable model

With ongoing information on the status of the system available, processes and their controls can be represented by numbers using the concept of state introduced in Chapter 4, Input into output/systems. As with material input, Eqs. (4.2) and (4.4), this will take two equations. Relating the status of the process directly to control input is the first, the (differential) *state* equation. A common version depending on the time is shown in vector/matrix format.

$$\frac{dx(t)}{dt} = Ax(t) + Bu(t) \qquad x(t_0) = x_0 \qquad (5.1)$$

where the symbols x and u, represent the state variables and their respective control inputs as vectors, and A and B are matrices of the coefficients that represent actual quantities in the real world. The second, the *output equation* is given by:

$$y(t) = Cx(t) + Du(t) \qquad (5.2)$$

where y and u are the output and input variable vectors and C and D the respective coefficient matrices. As discussed in Chapter 4, Input into output/ systems, here relevant is primarily the first term on the RHS.

Input–output model

Whether by necessity or choice, practicalities may call for evaluating control of a process without having or using information on its workings. Turning to *IO* models in what essentially is a black-box approach, control is again a forcing secondary input that somehow affects process output. Central to *IO* methods (see Chapter 4: Input into output/systems, Eq. 4.7) is the *transfer function* $H(s)$, the ratio relating output, $y(s)$, to input, that is, control, $u(s)$ as given in a more general multivariate version by Eq. (5.3a). To represent output under control, terms can be rearranged to obtain Eq. (5.3b).

$$H(s) = y(s)/u(s) \qquad (5.3a)$$

$$y(s) = H(s)u(s) \qquad (5.3b)$$

As noted in Chapter 4, Input into output/systems, in more detail, Eqs. (5.3a,b) represent differential equations that are already in the *Laplace domain*, the initial stage in a mathematical work up and standard in the *IO* model. Often not spelled out in the literature, it will be implicit through Eqs. (5.4a,b)–(5.7a,b) and the accompanying explanations. As innocent as they may look, Eqs. (5.3a,b) provide, nonetheless, a versatile tool for modeling control of process initial transient phases, the steady state, and stability problems, as well as other usages delineated below.

Both the *SV* and the *IO* approaches serve to address aspects of control at various organizational levels. Some of the major applications will be outlined in the examples that follow. Similar considerations may arise, for instance, in actively bioengineering control of cellular processes.

Prospects for control

In the discussion of technological control so far there were certain tacit assumptions. The first was, obviously, that the system is open to control so that the objectives of the process will be met. Another implied that given controls are effective, adequate, or perhaps even optimal. Also assumed was that the output indeed reflects all the detail of process behavior under control. These assumptions, especially with large complex processes, cannot always be taken for granted. Altogether, opting for control can raise critical questions regarding the system as a whole and its limits. Two problems are here of special interest.

Feasibility

Can a process be controlled? In terms of the *SV* model, can a working system in fact be forced from one state to another by control? In processes evolving with time, this directly relates to their dynamic properties. One would need realistic control inputs that would accomplish the transfer from one state to another in finite time. If such control exists, the process is *controllable*. In the presence of at least one process member that is refractory to applied control, a process is *uncontrollable* or *incompletely controllable*. Nevertheless, processes and controls may still be useful when *critical* process components are responsive.

Considered questions of *controllability*, quantitative testing can be the preferred approach. Turning to the *SV* model allows us to cover a wide range of circumstances, say, in having single or multiple inputs, different system components, and single or multiple outputs; however, before getting into any detail, a general observation is needed. It turns out that a significant factor in controllability is the quantitative makeup of the process and controls, as represented by the SV model—the state and control variable coefficients in matrices A and B. A common theme in checking for controllability is that control is feasible when the rows of these and related matrices used in the analysis are linearly independent. If linear independence (see Chapter 1: Change/differential equations, and Chapter 2: Consensus/linear algebra) was a precondition for solving equation systems, here it serves as a diagnostic for controllability even without solving model equations. Applying linear independence criteria such as those in Chapter 2, Consensus/linear algebra, is sufficient to obtain a definitive answer. As it were, controllability questions arise with respect to both the state of the system as well as its output, whether process behavior is constant or varying (autonomous and nonautonomous presented in Chapter 1: Change/differential equations). The problem also arises with discontinuous processes where it is approached in similar terms using recursion formulas.

To provide a gist of these methods two prototypes for testing state controllability with some generality follow. Leaving the detail to more specialized sources, the tests depend on conditions that have to be met by expressions related to the solution of state Eq. (5.1). (1) The *controllability matrix* is a (row) matrix combining, as if sequentially augmented, the coefficient matrix B and its products with consecutive powers of matrix A. When the number of rows of matrix A, is n, it has the form $[B] \, AB] \, A^2B] \, \ldots \,]$ $A^{(n-1)}B]$ where the tricolon partitions signify where multiplication is to take place first. The state equation has a solution only if the controllability matrix has full (row) rank (i.e., the rank equals the number of rows) is the matrix linearly independent, and the process is controllable. (2) The *controllability Gramian* is $\int_0^t (exp)^{A(t-\tau)} BB^T (exp) A^{T(t-\tau)} d\tau$, where $(exp)^{A(t-\tau)}$ is a matrix exponential, BB^T is a so-called *Gramian matrix*—hence the name—and the T superscript denotes the transpose of matrices A and B. It is an expression closely related to the solution of the second, the forcing term of the state equation. Beginning to compute by multiplying the two Gramian components, matrices B and B^T, generate a single matrix whose elements are the products of the respective elements of each of the two matrices. Further multiplying by the matrix exponentials, numbers, is straightforward and the integration applies to all matrix elements separately. Here what matters is that at the end of this protocol, the (square) matrix format is still retained. Recall from Chapter 2, Consensus/linear algebra, that such matrices have determinants (see there), also numbers. If not equal to zero the matrix is nonsingular and its rows are linearly independent. The point here is that when the controllability Gramian is subject to these operations and turns out to be nonsingular, the process is controllable. The actual computations can be extensive, but also provide an opening to computing the actual control.

Transparency

Part and parcel of managing or designing control systems, real-life information on process workings is commonly derived from some sort of output; however, especially in large and complex systems, whether the output indeed represents every critical system component can be an open question and a practical issue. For example, in processes with multiple inputs but a single output, a subprocess may affect the control via feedback and yet not be readily identified from the output. Whether it is possible to deduce the state of the system from its output is a question of *observability*.

Turning to quantitative testing, the specific question is whether given the output at the end of a process at state x_t and knowing the input one can determine state x_0 in finite time. It turns out that the mathematics of observability is closely related to that just described, addressing controllability. Retaining matrix A but replacing Eq. (5.1) input matrix B with Eq. (5.2) output matrix C, to ensure *complete* observability again requires linear independence, now of matrix *columns* (i.e., in tests that employ counterpart observability matrices and Gramians). As it were, mathematics in this section is used to also characterize other whole system features that go under detectability, reachability, stabilizability, among others, and out of the scope of this book.

Evaluating control options

Open or closed loop

Having decided that a (technological) system is to be put under extra control, the question may arise whether it should be an open or closed loop. One way to compare the two options is to ask whether, and how much, change in output of an open-loop system would result by installing also a closed-loop feedback mechanism. The *IO* model, which incidentally represents system components that are controllable and observable to begin with, can be handy. To compare the open and closed-loop options one could use the respective transfer functions. What will these transfer functions be? As the simplest example, there are univariate knowns in engineering as *single input single output* systems such as those in Fig. 5.1A and B. Here to be represented by appropriate scalar equations, keep in mind that what follows next would hold also for multivariate, multiple input and output, systems using matrix equations.

Whether taking the *open-loop* arrangement as given in Fig. 5.1A or its counterpart, the forward (upper) loop in Fig. 5.1B, the corresponding *open-loop transfer function*, H_{ol}, is given by y_{ol}/u_{ol} (see Eq. 5.3a). The *closed loop*, as shown in Fig. 5.1B, consists of a *forward branch* at the top, equivalent to the open-loop system and again representable by H_{ol}. Using the output of the open loop as input is the *feedback loop* (lower) whose output again enters the system. It has a *feedback transfer function*, H_{fb}, of its own. In real-life models, often more complex, specific transfer functions are first attached separately to participating loop components. The overall open and feedback loop transfer functions, H_{ol} and H_{fb}, are the products of those of their respective components.

Computing the total output of the closed-loop system with negative feedback as example, it will equal the output of the open loop less than that used by the feedback loop as input. In the form of Eq. (5.3b), this will lead to Eq. (5.4a). Minor manipulations of Eq. (5.4a) will lead to Eq. (5.4b):

$$y = H_{ol}u - H_{fb}y \qquad (5.4a)$$

$$y = \frac{H_{ol}u}{(1 + H_{ol}H_{fb})} \qquad (5.4b)$$

The overall closed-loop transfer function, H_{cl}, is then obtained by dividing both sides of Eq. (5.4b) by u and is given by Eq. (5.5):

$$H_{cl} = \frac{H_{ol}}{(1 + H_{ol}H_{fb})} \qquad (5.5)$$

Eq. (5.5) is the desired expression that provides for comparing open-loop and closed-loop operations. Evidently they are related through the term $1/(1 + H_{ol}H_{fb})$, a common presence in this type of analysis. Absent feedback, the denominator and the entire term equal 1, regenerating the original open-loop transfer function. In real-life technological systems, the term $(H_{ol}H_{fb})$ in the denominator may become much greater than the number 1 which, consequently, can be neglected. Thus when $(H_{ol} H_{fb}) >> 1$, H_{ol} will cancel out so that:

$$H_{cl} \cong \frac{1}{H_{fb}} \qquad (5.6)$$

Under these circumstances, the closed-loop system is sensitive to the feedback path, but not to the forward and environment-sensitive, open-loop component. Consequently, configured as part of a closed-loop system, the open-loop pathway need not by itself be very accurate or time invariant.

Sensitivity

Artificial processes are liable to operate under real-world conditions that depart from those taken into consideration in the original design or specified by a model. Similarly, and sometimes related, active interference of natural and human origins can cause process variability that is unacceptable. How this might affect processes, whether they would be stable and how, is a recurring theme in technological practice as has been seen already in Chapter 1, Change/differential equations, and Chapter 4, Input into output/ systems. Will control effectively eliminate the problems? These issues can be evaluated quantitatively. The methods are widely applicable also in dealing with other questions. Here, notably the modeling of cellular control in MCA as will be seen later on.

Evaluating off-target technological processes, as in monitoring stability, is based on the change in output brought about by a small alteration or *perturbation* of an appropriate input, for example, the change in the yield of a product of a chemical process resulting from a modest change in the supply of a precursor. The effectiveness of control is assessed by comparing process response with control to that without. A measure common in the *IO* model and already used above is the ratio of the change in the respective transfer functions, say univariate $\Delta H_{cont}/\Delta H_{proc}$, where Δ stands for the respective change; however, for technical reasons that are of no concern here, the evaluation in this form can be problematic.

Popular as an alternative is the *fractional* change due to perturbation known as *sensitivity*, S; that is, the ratio of the perturbation induced change in the control transfer function over its total value divided by its counterpart without control as shown by Eq. (5.7a). For continuous cases, a differential format, Eq. (5.7b), is common. Recalling from calculus that $1/x$ is the derivative of the natural logarithm, $\ln x$, Eq. (5.7b) is frequently taken one step further and given in the form on the extreme RHS.

$$S = \frac{(\Delta H_{cont}/H_{cont})}{(\Delta H_{proc}/H_{proc})} \tag{5.7a}$$

$$S = \frac{(\partial H_{cont}/H_{cont})}{(\partial H_{proc}/H_{proc})} = \frac{\partial \ln H_{cont}}{\partial \ln H_{proc}} \tag{5.7b}$$

Note that: (1) Sensitivity testing is also used for other purposes—for example, to evaluate the differences between closed and open-loop control, or to assess the contributions to process sensitivity of individual components of a control system such as those in Fig. 5.1. (2) Generalizing from single input or single output, scalar sensitivity equations to multivariate vector or matrix format is common.

Feedback

Attaching feedback to a control system by adding components and modifying the design evidently changes the structure and functions underlying a process. It also affects its quantitative makeup. That this could be significant could have been inferred from the evaluation of control options using *IO* model transfer functions in Eqs. (5.4a,b)−(5.6). With the *IO* description being based on the outer components of the input/system/output scheme, the *SV* model offers a complimentary and more detailed view that centers on where control is actually being applied, namely the system component of the scheme.

Contributing processes

To take a closer look, consider a possible *SV* model of the processes in a basic system such as that in Fig. 5.1B. The input term in this model, $u(t)$, combines a reference value, $r(t)$, and a feedback correction, as shown in Eq. (5.8a). The feedback correction depends on the state variable $x(t)$ and employs coefficients of its own as given by a *feedback gain* matrix, K. Combining the state and input variables into a standard state equation leads to Eq. (5.8b) and by minor further manipulation to the form on the extreme RHS. The output obeys Eq. (5.8c), that is, Eq. (5.2) without the second term. When feedback is negative, the feedback matrix K and the associated corrections, starting with Eq. (5.8a), have a minus sign.

$$u(t) = r(t) + Kx(t) \tag{5.8a}$$

$$\frac{dx(t)}{dt} = Ax(t) + Br(t) + BK\,x(t) = (A + BK)x(t) + Br(t) \tag{5.8b}$$

$$y(t) = Cx(t) \tag{5.8c}$$

As can be seen in Eq. (5.8b) (extreme RHS), introducing feedback by matrix K changes the values of the coefficients of the state variable $x(t)$ in the state equation. It thus also affects the eigenvalues of the coefficient matrix, which, as will be recalled from Chapter 1, Change/differential equations, determine the dynamics of a process, whether it would be stable, as well as global process behavior patterns. Eigenvalues are, thus, intimately related to control and, as do their counterparts, the *IO* model poles, they are significant factors in control design. Computing the eigenvalues of the combined matrix $(A + BK)$ by methods of linear algebra (see Chapter 2: Consensus/linear algebra) can be key to the design of appropriate control. Similarly, aiming to optimize control design (see below) may boil down to optimizing K.

Readers may have noticed something odd about what Eqs. (5.8a−c) describes. As it were, feedback in this system is taken into consideration before the system becomes active, as in Eq. (5.8a), but actually depends on the output which is available only after it was active. In other words, feedback implemented in Eq. (5.8b) depends on the as yet unavailable value of x in the output Eq. (5.8c). There is a similar situation in modeling with the *IO* in Eqs. (5.4a,b)−(5.6). Altogether a paradox not lost on modelers, the associated error is generally considered negligible and the models still valid.

Feedback implemented

As delivered in real life, feedback may take additional factors into consideration. With suggestive analogies in certain subcellular systems, notably in signal transduction, technological controllers largely employ three types of formula, alone or combined. One kind, (*P*), elicits a response that is proportional to the error signal, essentially reflecting the instantaneous state of the system, as expressed by Eq. (5.8a). A second type, (*I*), with biological parallels, is based on an integral of the error signal and thus in a sense represents system history. The third, (*D*), uses the derivative of the error signal, perhaps reflecting a future trend, and is commonly employed in conjunction with the previous two as a further correction. Popular in technological feedback systems are *proportional integral derivative (PID)* controllers combining an appropriate mix of all three types. As already intimated, obtaining sufficient data for feedback from the output is not always feasible for technical reasons. Instead, control data will often be generated based on a theoretical model of the process.

Optimizing controlled systems

Optimal performance is obviously desirable in a primary process but also in its control, or both combined—whether to obtain the best material output or to economize on resources, be the settings biological or technological.

An overview of process optimization with cellular applications is presented in Chapter 6, Best choices/optimization. Here noteworthy is that among the methods used in cellular modeling those employing the Pontryagin maximum principle discussed there, in fact, originated in optimizing technological control in the first place. Generally, implementing optimal solutions as such for various practical reasons may not be possible or even most desirable. *Neighboring* or *suboptimal* counterparts, not discussed here, may be preferable.

What to optimize? Among the more specific overall objectives common in optimizing technological control, some are potentially also of biological interest: (1) minimizing the time it takes to bring a system from one state to another, often a steady state; (2) keeping a process stably on track, that is, minimizing both the deviations from the target trajectory while a process is unfolding as well the error in the overall target value; and (3) minimizing the energy expenditures of a system.

Is what has been described here so far useful in exploring processes in cellular systems? Indeed, it is. The rest of this chapter will provide a glimpse at the unique infrastructure of a linear control theory—inspired approach to process control in biological cells.

Cellular self-regulation

A basic feature of cellular control is evident in metabolic multienzyme pathways where it has a long history of experimental and theoretical study, glycolysis especially. Pathways taken on their own are self-regulating, although ultimately they also depend on what takes place elsewhere in the cell, for example, the supply and demand for their metabolites or the actions' various effectors. Self-regulation employs internal feedback, with a role for the sensory, computational, and response capabilities of enzyme molecules that were pointed out earlier. Trying to understand control behavior in a sequence of reactions raises questions whether and which particular enzymes might have control over the entire throughput. If there are several, how would control be distributed? Will the distribution of control be affected by changing the conditions? What effects will various interventions have, whether by genetic engineering or by exposure to foreign agents? Where would intervention serve a certain purpose best? These are some among others that may be asked. Similar questions arise about the metabolites. As pointed out before, standard kinetic models recreate complex processes bottom-up and face significant practical and theoretical limitations (see especially Chapter 1: Change/differential equations, and Chapter 3: Alternative infrastructure/ series, numerical methods). Still the preferred modeling format, they do not necessarily provide all the desired answers. A change in venue could be in order.

Metabolic control analysis

Changing the outlook, there is the alternative option to confront complexity head on by probing for control behavior top-down. As a major contender in this field, *MCA* as it is now known goes back to the 1960s [1] and was formulated fully first in the 1970s [2,3]. While rooted in the conventional, custom, kinetic models of metabolic pathways it has an outlook and infrastructure of its own with applications also elsewhere, for example, investigating intracellular signaling or gene dosage effects. An *MCA* contemporary, *Biochemical Systems Theory* (initiated by Savageau) provides another major quantitative approach to control in cellular systems. Unlike *MCA* it retains the conventional kinetics point of view and infrastructure and is not discussed here. As it were, dealing with the same realities, there is also substantial overlap with *MCA* in the type of questions asked, methods employed, and information obtained.

Conceptually a black-box approach, *MCA* draws on the *IO* model and its methods. Unlike the detailed, standard biochemical descriptions, *SV* model related it does not employ a physical model, does not provide mechanistic detail, and in fact is often difficult to relate to commonly measured properties, notably individual enzyme rate parameters. Empirical, its results are valid primarily for a given set of conditions, but have already provided insights of a more general nature.

Perturbing the pathway and the membership

To evaluate control, *MCA* turns to the time-tested perturbation approach. Again, this calls for measuring the effect on an output of making a *small* change in input. At the core of this approach are the effects of perturbation on the throughput of the pathway, the *flux* of pathway metabolites, and on the activities of the individual enzymes. To be perturbed in the latter are the activities of member enzymes and the concentrations of metabolites and other effectors (additional cellular or foreign activity modifying compounds). Using the standard conditions of the steady state, data are collected in situ, be it an organism, tissue, cells, as well as extracts or equivalent conditions.

Quantitative models of these observations will again employ a fractional change format, in the general engineering sensitivity mold of Eq. (5.7b). Strictly speaking these models relate change in output to that in input within the *same* system rather than the output/input ratios of two *separate* systems of Eqs. (5.7a,b) and Laplace transformation is omitted altogether. It is a format of relating output to input used elsewhere, for instance in the economics of supply and demand. It is known there as *elasticity*, a term that here has a specific meaning.

Central to the *MCA* infrastructure are three types of coefficients. Of these, two *control coefficients* represent behavior measured on the entire system, that is, the entire *pathway*, and referred to as *systemic* or *in the large*.

Both express the effects of perturbing the activity of a member enzyme. Representing the control of the overall throughput is the *flux* control coefficient, $C_{E_i}^J$, where E_i stands for the activity of member enzyme i and J is the rate of producing the output, the final product of the entire system. Likewise, systemic and representing the effect of changing the activity of a particular enzyme, but on the concentration of any chosen metabolite is the *concentration* control coefficient for each metabolite, $C_{E_i}^{M_j}$, where M_i is its concentration. Used here strictly as labels with no other mathematical meaning, superscripts specify the dependent, subscripts the independent variables, for example, flux and enzyme activity, respectively. As before i or j are generic for any one particular out of a number of enzymes or metabolites in question. When it comes to numbers, the flux and control coefficients are given in the form $\partial J/J/\partial E/E$ and $\partial M/M/\partial E/E$, respectively, leading to:

$$C_{E_i}^J = \frac{\partial \ln J}{\partial \ln E_i} \tag{5.9a}$$

$$C_{E_i}^{M_j} = \frac{\partial \ln M_j}{\partial \ln E_i} \tag{5.9b}$$

The third coefficient expresses *local* or *in-the-small* properties of the *individual enzymes* measured in situ or in the test tube under presumably matching conditions. Representing the response of enzyme activity to small changes in the concentration of a metabolite, be it substrate, product, or for that matter an effector produced within the pathway is the *elasticity* coefficient, or simply the *elasticity*, ε. Again, in terms of fractional change, given the rate of the enzyme reaction, v, and the concentration of a metabolite (or effector) of the pathway, m, by definition

$$\varepsilon_{M_i}^{v_j} = \frac{\partial v_j/v_j}{\partial M_i/M_i} = \frac{\partial \ln v_j}{\partial \ln M_i} \tag{5.10}$$

where the super and subscripts are labels. Each pathway enzyme may have elasticities with respect to its matching, as well as other metabolites or effectors. Both of the control coefficients are, in fact, expressible in terms of the elasticities of system enzymes; sometimes it is actually possible to compute the former using the latter.

Additional parameters here noteworthy serve to accommodate more special needs of *MCA*. Most prominent are *response coefficients* such as R_Z^J, a measure of the dependence of a system variable, often the flux, J, on an *external* effector, Z, frequently an inhibitor. Again, expressed in terms of fractional change, $R_Z^J = \partial \ln J/\partial \ln Z$, where Z is the effector concentration.

Control sharing

Ultimately one would want to use these parameters to evaluate control in the entire pathway. To estimate the relative contributions to collective behavior of pathway components requires a value for the entire pathway collective behavior. Here it is given by the sum of the respective coefficients. Further work up involves calculating all combinations of enzymes and metabolites in a process and can be quite elaborate. Significant here is that a central role in these computations is given to the two control coefficient *summation theorems*. The sums of the flux and metabolite control coefficients over n pathway enzymes are given, respectively, by the following expressions:

$$\sum_{i=1}^{n} C_{E_i}^{J} = 1 \qquad (5.11a)$$

and

$$\sum_{i=1}^{n} C_{E_i}^{M_j} = 0 \qquad (5.11b)$$

Also employed in the computation are expressions of *connectivity* properties combining control coefficients and elasticities such as

$$\sum_{i=1}^{n} C_{E_i}^{J} \varepsilon_{M}^{i} = 0 \qquad (5.12a)$$

and

$$\sum_{i=1}^{n} C_{E_k}^{M_m} \varepsilon_{M}^{i} = - \delta_{mk} \qquad (5.12b)$$

Note that while Eq. (5.12a) is similar to Eqs. (5.11a,b) in representing all combinations of enzyme and metabolites, Eq. (5.12b) represents any one chosen particular enzyme and metabolite combination. δ_{mk} on the *RHS* of Eq. (5.12b) is the Kronecker delta, a popular mathematical device to ensure that only significant detail is included. Here it equals one when enzymes interact with their matching reactants and zero otherwise.

The share of each factor in affecting the entire pathway is the measure of control. With detail already covered by extensive literature, it turns out that control in self-regulating multienzyme systems is indeed reminiscent of that in the car traffic example at the beginning of this chapter. By the criteria of *MCA*, control of overall steady-state flow is shared among member enzymes, not necessarily all or equally, and is possibly shifting from one to another depending on the conditions. It can get more complicated in pathways that are branched and those with internal cycles, as can be seen in more advanced literature [4].

Keep in mind

Process control in both cellular and technological environments generally depends on immediate objectives that may include initiating and terminating a process, steering active components from one state to another, maintaining steady states, all of these while responding to random noise as well as active interference, internal and external.

Major cellular processes including metabolism and signaling are subject to internal control by pathway components and exhibit basic *self-regulation*. Major control, cellular and technological, is mediated by *specialized structures* that affect process activity directly. Two technological control prototypes are intimately tied to whether or not information is available about an ongoing process and have counterparts in cellular control operations. Technological *open-loop* configuration requires sufficient knowledge of the starting conditions of a process, its environment, and its projected behavior under control, and is preprogrammed. *Closed-loop* design, more effective but also more costly, is suited to processes that are less predictable, but provide ongoing information that is employed as *feedback* into the control system in real time.

Quantitative models are based on the input/system/output scheme in which control is a secondary input that governs a primary process. Using information on process workings, the system component, is the *SV* model with its state and output equations. Without process information, the black-box *IO* model centers on transfer functions. Both approaches employ differential equations in the time and the Laplace domains, respectively, with options for support by linear algebra and alternative computational methods.

Whether, what, and how technological controls should be operating is decided case by case. The feasibility of controlling the output of a process and its contributions, *controllability*, may be evaluated quantitatively in the *SV* format. It could depend on whether the state of all system components can be deduced from the output, a related question of *observability*. Whether control should actually be applied, open or closed loop, may be open to either or both the *SV* as well as the *IO* models. Opting for feedback control, the design can be guided by *SV eigenvalues* and *IO poles*. Evaluating control in *perturbed* processes relies on *fractional change* with *sensitivity* as the measure. Common objectives of technological control include minimizing the time it is applied, keeping a process stably on track, and minimizing cost.

MCA is an *IO* model informed black-box approach aiming to characterize the distribution of control and related process parameters among members of self-regulating cellular pathways. Top-down, it employs data on *system(ic)* response of the overall pathway *throughput* and *metabolite concentrations* to the *perturbation* of individual member *enzyme* activities and *metabolite* concentrations. Also needed is the *local* response of *individual* enzyme activities to perturbing metabolite concentrations. *MCA* infrastructure centers on *fractional change* in the variables of systemic *flux* and *concentration control coefficients* and the local single enzyme *elasticities*. Overall, pathway control properties, for example, its distribution, are evaluated using *summation* and *connectivity theorems*. Having originated in modeling multienzyme pathways, *MCA* infrastructure is being applied to other cellular processes, notably in signaling and genetic networks.

References

[1]   Higgins J. Analysis of sequential reactions. Ann. NY. Acad. Sci. 1963;108:305–21. Available from: https://doi.org/10.1111/j.1749-6632.1963.tb13382.x. PMID 13954410.

[2]   Kacser H, Burns JA. The control of flux. Biochem Soc Trans 1973;23:341–65.

[3]   Heinrich R, Rapoport TA. A linear steady-state treatment of enzymatic chains. General properties, control and effector strength. Eur J Biochem 1974;42(1):89–95.

[4]   Joy MP, Elston TC, Lane AN, Macdonald JM, Cascante M. Introduction to metabolic control analysis (MCA). In: Fan TM, Lane A, Higashi R, editors. The Handbook of Metabolomics. Methods in Pharmacology and Toxicology. Totowa, NJ: Humana Press; 2012.

Further reading

Auslander DM, Takahashi Y, Rabins MJ. Introducing systems and control. New York: McGraw-Hill Book Company; 1974.

Chen CT. Linear systems theory and design. 4th ed N.Y: Oxford University Press; 2013.

Dorf RC, Bishop RH. Modern control systems. 11th ed. Upper Saddle River, NJ: Pearson Prentice Hall; 2008.

Gopal M. New Delhi: New Age International Publishers, Reprint Modern control system theory. 2nd ed. New Delhi: Wiley Eastern Limited; 2010.

Heinrich R, Schuster S. The regulation of cellular systems. New York: Chapman & Hall/Kluwer; 1996. Now available from Springer.

Schulz AR. Enzyme kinetics. Cambridge: Cambridge University Press; 1994.

Chapter 6

Best choices/optimization

Context

To eat right, go places, communicate thoughts or, for that matter, to skin a cat there are various common ways. The same applies for biological success or survival. Major cellular functions can have at their disposal different processes and take place under a spectrum of biological conditions. In mobilizing energy this may entail regulating the internal environment, maintaining the integrity of the cellular machinery, or conserving the fidelity of the genetic program, among others. External conditions can vary, say, temperature or pH, and there can be limitations on essentials such as nutrients or water. Is one option *best*? If so, *which* is it?

Taking a specific route expresses a *choice* of one particular option among a number of available ones. Of those available, there may be one option (sometimes more) that is the *most favorable*. Call them *optimal*. Being optimal or *optimality* is a major theme in understanding nature and, obviously, in managing the artificial world. It has a long history in higher-level biology and is now increasingly documented in living cells as well. How can cellular processes be optimal? What decisions might be involved? Two idealized scenarios can illustrate much of what is known about optimality in cellular processes.

In the first scenario a cellular function has at its disposal separate and independent alternative processes to choose from. To obtain energy, for example, cells metabolize mainly carbohydrates, but can also use fats or amino acids. Some of these processes, in turn, can take place along different pathways. The benefits and costs of each option differ, for example, in energy yield, in the metabolic cost of synthesizing and maintaining the various pathway machineries. And there may be important additional factors—effects on other cellular functions, the cells' physiological state, a specific biological role, among others. What would be the best choice?

In the second scenario a cellular function depends on a combination of processes, some incompatible. To make the best of this situation the cell needs to manage their relative contributions. In a case that will be revisited later on, overall cellular growth, that is, cell division with subsequent growth processes, depends on materials generated by photosynthesis. In these cells, growth could take place any time during a daily cycle, but photosynthesis is

Collective Behavior in Systems Biology. DOI: https://doi.org/10.1016/B978-0-12-817128-8.00006-7

151

restricted to daylight only and is inhibited by growth processes. Is there a time allocation of the two processes that will make growth optimal? Seen in more general terms the first scenario represents a choice of the best single or combination of compatible pathways; the second scenario seeks the best compromise between incompatible processes. In reality, however, these distinctions are often not clear-cut. One is obviously interested in finding process optima in any event, but knowing that a given process is *not* optimal can be valuable as well.

Evaluating optimality often turns to quantitative criteria, *how much* is best. In this chapter processes are taken to be optimal when their *value* corresponds to an *optimum* (or *optima*), that is, a *maximum* or a *minimum* among possible values. In other words, values of the describing expression can be *extreme*, a synonymous term that in its various derivations is also common in the literature. Moreover, science maintains that quantitative optimality is widespread in nature; the majority of scientific theories have it as a default assumption. Well-known examples in the real world range from light minimizing travel time when propagating through a medium, to industrial production maximizing output relative to input, to the individual or nation prospering by properly balancing consumption and savings. In higher-level biology, this is realized by organisms that time activities in register with nature's cycles, balance resources between growth and reproduction, and surviving extreme environments. Examples on a whole-organism level are human gait optimizing force and energy usage or a peacock displaying maximizing reproductive opportunities.

Optimality has, in fact, been demonstrated in whole cells and in their lower-level systems. Unicellular algae that allocate photosynthetic products optimally will be discussed later on in some detail. Glycolytic pathway reactions with ATP appear in models to have been placed so as to maximize energy capture. The timings and levels of synthesis of consecutive enzymes in bacterial amino acid biosynthesis pathways are optimal under the constraints. There are enzymes that are most active at body temperature. Ribosomal protein synthesis in response to change in nutrient conditions in *Escherichia coli* is likely optimal by theoretical criteria. Practical applications of cellular optima already include adjusting therapeutic timing of antiviral or anticancer treatments and maximizing yields of biotechnological fermentation processes, no less also genetic engineering in biotechnology and agriculture that balances human needs for useful, but deleterious, foreign genes with overall yield and commercial value. As it were, optimality properties provide theoretical tools for predicting metabolic behavior, charting cell shape development, and establishing sequence homologies between nucleic acid or protein sequences.

Everything considered, one might expect optimality to be extensively documented in cellular processes in their various manifestations. Except for relatively few cases (such as the above) it still largely remains to be established which particular cellular processes are indeed optimal or near optimal—when, by what criteria, and for what purpose. It may not even be clear, offhand, what an optimum of a particular process would be, whether or not one has been already attained, or whether it would necessarily be advantageous to the cell on the whole. And at the whole-cell level questions arise. Do cells follow any general patterns in optimizing their processes? How much in metabolic resources should cells invest in signaling and control devices or in the machineries responsible for correcting or eliminating faulty cell constituents? When, and why, should cells cope with the latter by choosing death instead? Does, or should, Darwinian evolution aim for maximal adaptation to particular conditions or perhaps for suboptimal, but more environmentally flexible balance?

The practice of *optimization*, here the search for real-world quantitative optima, has a long history and by now a rich infrastructure. That the circle has the largest area of all shapes that a rope may enclose was known already in ancient Greece. Modern mathematical optimization, beginning in the 18th century, was originally concerned with optimizing the dynamics of bodies falling and moving forward at the same time as well as other process properties. Having since developed in various directions, it currently extends over a wide range of applications in science and technology. As befits real-world diversity and in line with much other modeling in this book, optimization procedures are customized to a combination of factors. The particular system (types of processes, possible limitations), the objectives of the optimization, information already available, and to the mathematics currently at hand.

Present and likely future approaches to optimization in systems biology are diverse. Infrastructure for two of these, mainstays of formal optimization, will be featured here first. *Classical* optimization procedures typically hinge on having process information sufficient for constructing a unique mathematical statement similar to process models elsewhere in this book. Optima (i.e., maxima and minima) will be identified by methods familiar from the calculus. An example in some detail will illustrate how cellular growth can be maximized by finding the best balance between two independent and constant, but incompatible, contributing processes. *Variational* optimization, the second approach, is tailored to the quite common circumstances in which initial information is insufficient for defining a unique optimum, but does provide for a range of candidates to choose from. Using a strategy that aims directly for an optimal value of an overall process outcome, the variational approach will here come in two versions. The *calculus of variations*, the original scheme providing basic concepts as well as practical tools will be introduced first. Research applications already include optimizing industrial biomass production and charting cell shape development. Taking it further will be a *Hamiltonian* format (a third approach) equipped with a *(Pontryagin) maximum principle*. Originally created to optimize technological combinations of processes and strongly interacting controls, its methods are now applicable also elsewhere—here notably in increasing optimality studies of cellular growth and resource allocation.

A range of practical optimization needs is served by two computer-oriented approaches that will occupy the latter part of this chapter. *Linear programming (LP)* combines algebra with a geometrical description. It is already used widely in engineering and various social sciences. It is also gaining ground in systems biology, for example in optimizing and characterizing metabolic behavior, as will be seen. *Dynamic programming* offers a general strategy for efficient handling of optimization procedures that require extensive computation. Wide ranging over various fields, it will here be illustrated as applied to searches for protein sequence homologies. Inevitably there will be the occasional mathematical detail throughout this chapter. The main focus, nevertheless, will be on more general features of quantitative optimization procedures, materials that are accessible to all readers. Still other models of process optimization that are closely tied to network organizational structure and relate to graph theory will be discussed in Chapter 8, Organized behavior/network descriptors.

A formula for the optimum

Optimizing cellular processes traditionally aims for optimal values of familiar chemical parameters, say, reaction rates or amounts of cellular components. More recently optimization has been extended also to *mechanical* properties, notably elasticity related, in cell elongation and division. Having chosen what to optimize, one would next set up a model, once again employing mathematical functions that represent that particular property assembled into equations. In many cases it is possible to optimize directly process models such as those in previous chapters. As will be seen, however, some cases call for other special expressions. In any event what is to be optimized will be known here as the *performance measure/index*. Other terms include *objective function* as well as *fitness-*, *profit-*, or *cost function*, also *figure of merit*, terms that originally related to particular context—nature/science, technology, business, and so forth.

An initial, most general, statement of an optimization problem could simply read:

$$\text{Optimize } P = f(x) \tag{6.1}$$

where P is the performance measure as given by the mathematical function (s) $f(x)$. An optimum would then correspond to a maximum or minimum value of a performance measure. The value will be obtained solving equations of the performance measure formally or by alternative methods. While the emphasis here will be on continuous models to begin with, discontinuous versions are open to optimization as well. Note, however, that due to the mathematics, not every performance measure necessarily has a computable optimum.

Moreover, even if one or more is found, a different formula could lead to optima that are as good or even better (or worse). Given this point of departure, how will optimizing cellular processes play out?

The classical way

For one, *classical* optimization turns no further than to some most elementary notions of maxima and minima. Familiar from basic calculus, they were reviewed briefly in Chapter 1, Change/differential equations. When a process is at optimum (i.e., maximum or minimum) the *first derivative* of a function representing the problem, say, a performance measure, equals zero. It has a maximum value when the *second derivative* is smaller than zero (i.e., with a negative sign) and a minimum value when it is bigger (i.e., positive).

A real-world cellular example can be found in the optimization problem of the second scenario above. The unicellular alga *Chlamydomonas reinhardtii* is an autotroph in which photosynthetic product has two major destinations—the synthesis and maintenance of machinery for cell *division* and *growth* (largely protein) and the accumulation of *storage* materials (carbohydrates) for eventual use in these processes. Growth may also be supported by materials from cellular breakdown. As it were, growth processes compete with photosynthesis. Cells could manage by separating the processes in time. Do the cells during a normal 24-hour cycle, in fact, follow a regime that optimally balances the processes and maximizes growth?

Employing classical optimization, the Cohen−Parnas model [1] addresses a number of related questions. A detailed model of contributing processes in the original, the highlights that follow here in more simple terms should provide a taste of this approach. The objective of the procedure is to determine the time allocated exclusively to photosynthesis that will maximize overall growth. Photosynthesis time is reflected in the amount of total materials synthesized, a measure of total photosynthesis, and the portion of these materials to be stored. It is directly related to the *ratio*, S, between the weights of the storage and the total photosynthetic materials at the end of daylight. Of all possible ratios, there will be one that will optimize the growth. Using technological terminology, the performance measure to be optimized is here, appropriately, called the *profit* function, π, a function of S, or $\pi(S)$. The optimal time allocation will be given by the ratio S that corresponds to an optimized π.

What will optimize the growth? This is up to the modelers. The present model takes into consideration factors in addition to those mentioned so far. (1) Cell division takes place in the dark. Together with cell maintenance, this will require usage of storage material. If not enough of the latter is available, cells will be broken down to provide energy and precursors. (2) Shutting growth off during photosynthesis is a loss of opportunity to grow. Taken together, the performance measure, $\pi(S)$, is defined as the difference between a *gain* function, $T(S)$, and a *loss* function, $L(S)$, that is, $\pi(S) = T(S) - L(S)$. The gain, T, is taken to be the savings in cellular breakdown that corresponds to a minimally adequate supply of storage material. Loss, L, is taken to be the loss in growth when biosynthesis is channeled to storage.

Biologically, it is likely that various salient cellular processes will differ whether the amounts of storage materials are above or below the putative optimum. Therefore, so would $\pi(S) = T(S) - L(S)$. Under special growth conditions, equations could be simplified and solved analytically to provide optimal values of the profit function and the optimal product ratio and, thus, the optimal photosynthesis time. In the more general case, obtaining an optimum by solving the equations at hand formally was not practical. The optimum S_{opt}, was therefore determined from the putative position of the derivative of $\pi(S)$, estimated from mathematical properties of the processes and the profit function on both sides of the optimum. Experimental data were consistent with algal cells indeed following an optimal regime according to this model.

Note, incidentally, that π is here the dependent variable (on the y axis), and S is the independent variable (on the x-axis). Common calculus optimization problems ultimately aim for the optimal dependent variable. Here it is used to identify an optimal value of an independent variable.

Altogether, optimization models such as that can be created when there is what is considered *complete knowledge* of the system. In other words, the performance measure could be represented by *unique* and *initially specifiable* mathematical functions. In this particular case this was possible because contributing processes took place at a constant rate, essentially independently, and there were sufficient data available for modeling. Also noteworthy is that the mathematical solutions in classical optimization are *local*. Although here there was only one optimal point, in other cases there could be additional optimal points somewhere else along the curve. Among those it is likely that one is *global*, that is, with largest or smallest value (see Fig. 1.2). Finding the global optimum of complex processes can become a separate and demanding problem whether using analytical methods or computer algorithms.

Constraints on reality

Nature and various human-related factors evidently impose *constraints*, limits on real-world processes. A cellular process may be constrained, for example, by the natural rates of its reactions, the extent of driving forces determined by distance from chemical equilibria, or its susceptibility to extreme temperatures. Limits on cellular processes could also be imposed by other factors such as the availability of precursors or tolerance to deleterious byproducts. Implicit in describing photosynthetic resource allocation in *C. reinhardtii* above, among other processes in this book so far, was that the processes occurred within the permitted range and constraints and, therefore, could be neglected. This, however, is often not the case; many real-world optimization procedures are in fact intentionally designed to deal with constrained situations.

To identify so-called *constrained optima* quantitatively, constraints on the values of the performance measure, c_i, can be represented by auxiliary functions of the same independent variable, $g(x)$ or by constants. The respective equations could take the form of an *equality*:

$$c_1 = g(x) \tag{6.2}$$

or an *inequality* such as

$$c_2 \leq g(x) \tag{6.3}$$

Seemingly a subtle difference between Eqs. (6.2) and (6.3), it nevertheless may cause optimization procedures to take dissimilar routes.

In a widely used approach to *equality* constrained optimization, the solution process is initiated by combining the function to be optimized, f, with that representing the constraint, g (the argument x conveniently omitted). This creates an *augmented* performance measure, f_a, also known as a *Lagrangian, L*, as shown in:

$$L = f + \lambda g \equiv f_a, \tag{6.4a}$$

where

$$\lambda = \frac{-\partial f / \partial x}{\partial g / \partial x} \tag{6.4b}$$

where λ in Eq. (6.4b) is a *Lagrange multiplier*—an expression that relates change in the performance measure to change in the constraints. Further work up may then proceed in the unconstrained manner. It is convenient, for example, when representing multiple constraints by separate equations, each with a Lagrange multiplier of its own. Optimizing with equality constraints often employs a completely different approach (see LP later on). *Inequality* constraints commonly are more difficult, but sometimes can be circumvented, for example, by expressing the inequality as part of an equality constraint. With inequality (3) above, this might be possible by substituting with

$$c_2 = g(x) + (d)^2 \tag{6.5}$$

where d is a *slack variable* that represents at each point the difference between the values of the inequality and that of the equality.

Note also that: (1) Examples of procedures for optimizing cellular and other real-world processes with explicit constraints can be found later in this chapter. See the maximum principle method in modeling control, Eq. (6.8), and LP in flux balance analysis (FBA), Eqs. (6.9a)–(6.9c) and (6.10a,b). (2) Lagrange multipliers can also be used in special types of optimization problems.

Variational choices

A major approach to optimization in exploring nature and in managing the artificial world is offered by *variational* methods (name will be clarified below). A classic example that the optimality these aim for is evident in the real world in that light that travels between two points, propagating along a path taking the shortest time (Fermat's principle). With a unique angle on optimization, the variational approach has since its inception in the 18th century been widely employed in science, engineering, economics, and finance as well as various mathematics. Applied to biological cells, variational optimization is used with biochemical processes such as the cellular product partitioning problem, the timing of cellular response to anticancer or antiviral agents, as well as optimizing industrial bioreactor output, among others. It also serves in modeling cellular behavior based on mechanical properties. Cell-tip elongation or cell-wall division, for example, can be characterized, assuming that cell wall elastic properties are optimal under minimal energy conditions.

Turning to optimization by variational methods can be a matter of choice as well as necessity. They are favored when they offer direct access to the actual values of often-desired optima, say, the maximum yield of a product or the minimum cost of a process. Necessity arises when knowledge of the system, as is often the case, is insufficient for unique performance measures such as that in the classical optimization example above—also, when other methods become mathematically problematic or are unsuitable to begin with. Still, lowering the need for information that goes initially into an optimization model widens the scope of the overall approach. Opportunities of potential interest in systems biology that variational approaches offer beyond those in classic optimization include optimizing processes with varying rates and such where a number of (possibly antagonistic) processes interact and over time could change their relative contributions, that is, with or without constraints. Another example is the capability of some variational methods to deal with processes with irregularities such as trajectories with corners and stepwise jumps, among other discontinuities. So far known mainly in technological processes, there could be counterparts in cellular behaviors as well.

To see what the calculus of variations is more directly concerned with, consider first a nonbiological example—optimizing the distance between two points. Walking down a street, without much ado the shortest most practical, but here least interesting, is the straight line. On the other hand, hiking through a valley between two mountain tops, it is not immediately obvious which route will be the shortest, take the least time, minimize overall effort, or maximize pleasure of the scenery. The question here is often not only which would be the preferred route, but also by how much; say, how short the shortest time or distance will be. Establishing where and when it will be best to keep right or left, maintain or change altitude, make more or less of an effort, and so forth, will depend on the particular circumstances and involve a choice between various candidate trajectories, each with its own description.

In practice, the variational approach begins with defining the optimum as a *sum total* value associated with the performance measure. The performance measure is the *(definite) integral* of the functions chosen to represent the problem. To be (literally and formally) *admissible* as performance measures, functions are required to have derivatives and integrals needed for a solution process that employs differential equation methodology. The optimal value itself is unknown and the performance measure in effect represents a *range* of plausible options, but lacks sufficient detail to compute an actual value. That is what you might indeed expect when, unlike the classic optimization case above, the initial information on the process is insufficient for a unique performance measure.

Optimal curves can be identified specifically, nevertheless, with the variational approach. Two versions with cellular applications that differ in their infrastructure and are used depending on the problem will be discussed here. The first, the calculus of variations, a mathematical discipline in its own right, provides basic concepts together with tools for immediate use. The other, related but in a Hamiltonian format, is versatile and increasingly finding its way into cellular research.

Calculus of variations

Formally, optimizing many variational problems can be represented by:

$$\text{Optimize } J = \int_{x0}^{x1} F(y(x), y'(x), x)\, dx \tag{6.6}$$

where J, the performance measure, is a definite integral of F which is not an ordinary, completely defined function, as is common in calculus (see Chapter 1: Change/differential equations). F here stands rather for a (super) function, technically a *functional*, of all functions $y(x)$, which offhand might generate an optimal value of the integral and that, by definition, are required to have a derivative $y'(x)$. As mentioned above, which function among the admissible is optimal depends on the value obtained from the integral. When optimizing biomass production, for example, as in many cases before, x could correspond to the time, t, and y might stand for a function that describes a value associated with the process, say, the ongoing amount of generated biomass. Constraints, frequent and defining features in many problems, can be represented by special terms. These are omitted here for convenience, but will be prominent later on. Integrals in Eq. (6.6) are to be taken between endpoints x_0 and x_1. The end points in the calculus of variations can be fixed, as they are here, or varying ("free"), as described by separate equations. Essentially partial differential equation *(PDE)* boundary conditions (see Chapter 1: Change/differential equations), they generally play a significant role in the work up. Altogether, what is described so far for single mathematical functions in scalar format applies also in many real-world problems that involve optimizing multiple functions in vector/matrix format.

The objective of the optimization procedure is to obtain from Eq. (6.6) a formula that will provide the one optimal value. That is, a function, $y^*(x)$, whose y value trajectory optimizes the performance measure. A somewhat elaborate process, a few highlights will provide the gist. Readers not interested in detail can skip this explanation and rejoin at Eq. (6.7).

Consider identifying a minimizing performance measure. As depicted in Fig. 6.1 there is an *upper* curve that can represent any arbitrary admissible function. Assuming that a minimizing function exists somewhere, it is not known, offhand, whether this curve is the optimal, how it can be established, and, if it is not optimal, how the truly optimal can be found. To answer these questions, evident to the reader but unbeknown to the upper curve, there is in Fig. 6.1 a *lower* curve that turns out to be the truly minimizing.

To make progress it will be useful to consider the problem from the point of view of the minimizing, lower curve. In that view, the upper curve is the sum of the values of the lower curve and the increment between the two, striated in Fig. 6.1. The difference in area under the curves, corresponding to the difference in integrals of the representative functions is a *(total) variation*, ΔJ. Assumed small, the increment is in a sense a perturbation of the lower curve. Mathematically the perturbation can be defined by a parameter, a small number, ε, and a function of x, $\eta(x)$, whose product, $\varepsilon\eta(x)$, is called a *variation*. Given the right make up, variations can be differentiated using the rules of calculus providing *first* and *second* variations, δJ and $\delta^2 J$. Analogs of first and second derivatives of the calculus, these can be used similarly in optimization procedures. To decide whether a curve is optimal takes a first variation that equals zero (i.e., $\delta J = 0$). Establishing that it corresponds to a maximum or a minimum of the performance measure takes that the second

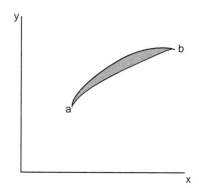

FIGURE 6.1   Admissible and optimal curves.

variation, $\delta^2 J$, is not equal zero (i.e., $\delta^2 J \neq 0$) but has, respectively, a negative or positive real value (i.e., $\delta^2 J < 0$ or $\delta^2 J > 0$). Altogether this is a calculus in its own right and the origin of the name of the entire approach.

Turning once again to Fig. 6.1 the optimal curve is still not known, but can reasonably be taken to have no increment attached. The given admissible curve, the upper in the figure, most likely would, as described by Eq. (6.6) and also including an increment term, $\varepsilon \eta(x)$. The parameter, ε is here key because the computation aims for the optimal curve to be given when ε equals zero. This will be computed beginning by differentiating the combined integral with respect to ε. Further setting this derivative to $\varepsilon = 0$ and integrating does not provide the optimizing function as yet. But, and here the major point, it establishes a *condition* for optimality.

The condition for optimality is met by satisfying the *Euler−Lagrange* equation:

$$\partial F / \partial y^*(x) - d(\partial F / \partial y^*(x)')/dx = 0 \qquad (6.7)$$

The Euler−Lagrange equation is a partial differential equation where y^* (x) is now the optimizing function and the partial derivatives, symbol ∂, are respectively those of F with respect to $y^*(x)$, and of F with respect to $y^*(x)'$, however derivatized with respect to x. The ultimate objective of the procedure, the desired minimizing formula $y^*(x)$ can generally be determined from the Euler−Lagrange equation by differential equation solving methods. With real-world problems, formal solutions not being practical, computations often turn to numerical methods.

The approach so far has been tailored to the simplest case in which both y and x are givens at both termini, say, the beginning and end of a process, as seen in Fig. 6.1. Real-world optimization problems often involve other contingencies. As already indicated above there are, for example, processes whose optimization may allow for either or both values at the end to be flexible and to be computed as part of the procedure. Specified by separate expressions, they can be accommodated by the Euler−Lagrange approach to optimization, but will not be covered here. Optimization of constrained processes, augmenting the performance measure with respective terms can follow similar lines as well.

Methods of the calculus of variations are established in the technological environment and have been successful already in optimizing certain cellular and biotechnological processes. Success, however, can be costly and not always feasible. Especially notorious is optimizing active control, technological and biological. What follows is a related variational approach that addresses some of the issues and extends the range of variational optimization. However, even in the barebones, highly simplified version, it still is demanding. It is included here, being both increasingly applied in cellular modeling as well as an important addition to the discussion of control in

Chapter 5, Managing processes/control. Intended for motivated readers, it can be skipped by others who are invited to rejoin the main storyline at the coda below.

A Hamiltonian format with a maximum principle

Optimizing control can be a major technological concern. Typically, one seeks to optimize an overall outcome of constantly interacting processes such as a primary, say, production process with a secondary process of control. Constraints on the primary process can become a significant modeling issue. Controls may be subject to natural, technological, or economic limits as well, often represented by inconvenient inequalities. Systems such as these also can develop irregularities that, among other factors, are a challenge to the optimization protocols described so far.

There are, however, workable alternatives. Still building on the calculus of variations and using some of its methods, they, at the same time, also draw on the mathematics used with optimal systems in physics. An infrastructure that is already widely employed in engineering, economics, finance, and elsewhere, there are also wide-ranging applications in cell biology. From attempts to explain how somatic mutations prepare antibodies for an effective immune response, exploring what enzyme concentrations would be optimal in metabolic processes, determining how the production of bioengineered foreign agents should be balanced with competing native growth processes, to optimizing drug therapies, among various others.

Underlying is a technological optimization prototype originally intended to identify controls that will optimize the overall return of a process under various real-world constraints. In this so-called *control problem* prototype, processes are represented as a progression of *states*, x, as in the *SV* model in Chapter 4, Input into output/ systems, say as given by the concentration of a cell constituent. Change in the state as a function of time, $x(t)$, as well as in the status of the control, $u(t)$, will be given again by ordinary differential rate equations such as those described in Chapter 1, Change/differential equations. Process and control, however, are mutually dependent, which will be represented by appropriate, time-dependent Lagrange multipliers, $\lambda(t)$, (see Chapter 3: Alternative infrastructure/series, numerical methods), here also known as *costate* variables.

As an elementary prototype, the literature often turns first to an optimization procedure that aims for an optimal control, u^*, that will optimize the overall performance of an entire system when, in formal terms, it brings a process from one state to another state (see Chapter 5: Managing processes/ control). Once again in the general optimization format of Eq. (6.6), it can initially be expressed by a performance measure J, as in:

$$\text{Optimize } J = \int_{t_0}^{t_1} f(x(t), u(t), \lambda(t), t) \, dt \qquad (6.8)$$

where time, t, is the independent variable and the endpoints, t_0 and t_1, represent the fixed beginning and, as yet, free ending times of the process. In many a real-world optimization, the performance measure also features an extra "penalty" term describing the primary process before control. Being of no special interest here, it is omitted.

Instead of using for the computations the mathematical functions of the performance measure, however, these are replaced by *Hamiltonian function (s)*. Not shown, here what matters is that, when derivatized, the Hamiltonian functions would regenerate the original (Lagrangian) functions and, thus, are Legendre transforms of the originals. Counterpart to integrating the original functions, making this transition has various advantages. Here notable is the disappearance from the formula of the first derivative y' and its requirement for continuity. This allows to include discontinuous processes and controls. It also puts on equal footing the differential state variable and the algebraic constraints, which now can be made time-dependent differentials, as well, in effect, becoming costate variables. Computing optimal process and control values that optimize the entire control system takes meeting variational optimization criteria such as those below combined with differential equation solving methods.

A major point is that protocols that determine the actual values associated with an optimized process differ, depending on the available controls. There is one version for controls without constraints and another based on the maximum principle below when they are constrained. In the latter case, it is specified, for example, by $0 \leq u \leq U$ or similarly as $u \leq U$, where U is a maximally allowed value of the control variable, u.

Optimizing *unconstrained* control, u, is handled using essentially the same calculus of variations procedures described above. Key to the procedure is that the condition for control being optimal is $\partial H(x, u, \lambda, t)/\partial u = 0$. The condition requires obeying a partial differential equation that equals zero and consists of the derivatives with respect to u of those terms in which it appears in the Hamiltonian, H. Solving the equation will determine the actual trajectory of the optimal control, u^*. Optimal state and costate variables, x^* and λ^*, are given by related equations that are first obtained and then solved similarly. These relationships hold also when control is constrained, but is still "unsaturated," that is, away from its constrained value.

When controls are effectively *constrained* to a certain range of activity the computations follow similar lines with the optimal control variable, u^*, now obeying a *maximum principle* formulated in the mid-20th century by Pontryagin and associates. For historical reasons it is often referred to as *minimum principle* because it is used more commonly in minimizing rather than maximizing performance measures, incidentally also more frequent in real-world variational applications. In this scenario the criteria control has to apply also to control trajectories that harbor irregularities such as corners and jumps. Altogether it is stipulated that within the allowed range, the fully constrained control will be maintained—as the name of the principle implies—at its minimal or maximal optimal level constantly and throughout a process' entire lifetime.

Putting the maximum principle to work, certain optimal technological controls follow what is known as a *bang–bang* regime. Controls *switch* abruptly between two extreme values, for example being maximally *on* and completely *off*, the timings of the switches can be calculated if needed. There is increasing evidence that certain cellular systems indeed behave as this description predicts.

Coda

All told, and returning once again to the main variational storyline, continuous mathematics versions of variational methods such as those in this chapter are known as *indirect* methods. Rigorous and in their original format, they are often detailed, difficult to handle, and solved numerically. To meet practical needs there are workable *direct* methods that optimize instead discrete, algebraic expressions. The integral in Eq. (6.6) may for example be replaced initially by the sum $\sum_{i=1}^{n} a_i \phi_i(x)$ of a sequence of *algebraic functions* of the same variables, $\varphi_i(x)$, chosen depending on the problem at hand, where a_i are coefficients. In effect approximations, format and motivation of this approach in many ways resemble those of alternative methods in Chapter 3, Alternative infrastructure/series, numerical methods. Still another approach to optimization are *search methods*. In this old strategy one starts with a reasonable representation and gradually improves the outcome by iterating a set of mathematical operations until there is an acceptable outcome.

Mathematical programming

Bearing in mind what today are somewhat misleading names, two widely used approaches to optimization occupy the rest of this chapter. Both are computer oriented as the term *programming* in the subheading suggests. The term programming, however, actually predates large-scale electronic computation and originally meant selecting the optimal among available options. It will also here be understood as a synonym for optimization when it refers to the two methods (linear and dynamic), otherwise unrelated, that follow.

Linear programming

LP is a versatile approach used widely in engineering and economics. There are also applications to cellular processes. As it were, especially in metabolism, there is already sufficient information to predict overall process behaviors in networks (Chapter 8: Organized behavior/network descriptors) and pathways as well as how these might be affected by their members and change in their makeup. It would be useful, for example, to identify in whole cells sequences of metabolic steps that optimize overall growth, biomass production, or the synthesis of particular components, say ATP, penicillin, or alcohol. It could also be helpful to know whether, and which, components might dominate process collective behavior and predict how it might respond to factors such as nutrients, particular effectors, the environment, or genetic change. Offhand, one would like to take into account process dynamics and process behaviors that may change over time, possible, for example, using the Pontryagin principle. The latter, however, is currently impractical, but there are opportunities elsewhere.

FBA offers procedures for optimizing overall processes as well as deal with the effects of specific factors such as those just mentioned. *FBA* essentially foregoes the time element by focusing on the steady state which, as readers will recall, is a quasipermanent feature and usually the predominant phase of ongoing processes. The steady state is also well-defined experimentally and mathematically and is useful as a defining constraint as will be seen below.

To begin the optimization of biomass production, for example, each metabolic step is represented by a linear ordinary differential equation *(ODE)* system (see Eqs. (1.8) and (1.18a,b)). However, the point here is that to actually identify optima, *FBA* turns to LP. Replacing the original *ODE*s with linear algebraic counterparts, *LP* is naturally computer-friendly and with increasing numbers of variables is often the practical way to go. Even when *LP* is applied to diverse cellular or other real-world, complex, processes with their various descriptions, these can be brought into a common initial format.

To begin with a most elementary, noncellular example, and in the footsteps of Eqs. (6.1)−(6.3), *LP* may be recruited to maximize a performance measure *P* such as:

$$P = c_1 x_1 + c_2 x_2 \tag{6.9a}$$

subject to the constraints

$$c_3 x_1 + c_4 x_2 \leq b_1 \tag{6.9b}$$

$$c_5 x_1 + c_6 x_2 \leq b_2 \tag{6.9c}$$

Here c_{1-6} and $b_{1,2}$ are constants and x_1 and x_2 are the variables to be solved for. That performance measures are *maximized* is standard in *LP*; minimizing problems are readily transformed to a maximizing format first. Also seen, constraints with *b* as upper limit (Eqs. 6.9b and 6.9c) are allowed as both equalities and inequalities. As with optimization procedures before, there may be *LP* models that do not have an optimum to begin with.

Real life optimizations with *LP* routinely involve multiple variables; problems are presented in vector/matrix form. Eq. (6.10a) represents the general case corresponding to the (scalar) Eqs. (6.9a)–(6.9c). Evidently, as seen in Eq. (6.10b), maximization of biomass/growth by *FBA* is in the same mold.

$$\begin{aligned} \text{Maximize} \quad & P = c^T x \\ \text{subject to:} \quad & Ax \leq b \\ & x \geq 0 \end{aligned} \tag{6.10a}$$

$$\begin{aligned} \text{Maximize} \quad & P = c^T v \\ \text{subject to:} \quad & Nv = 0 \\ & 0 \leq v_i \leq \beta_i \end{aligned} \tag{6.10b}$$

Beginning with Eq. (6.10a), the symbol c^T is a row (transposed) vector (see Chapter 2: Consensus/linear algebra) of coefficients c, such as those in Eq. (6.9a). x and b in Eq. (6.10a) are column vectors of their counterparts in Eq. (6.9a,b,c), and A is a matrix of the coefficients of the constraints, say, Eqs. (6.9b) and (6.9c). The bottom row expresses a common stipulation, namely that the variables assume positive values. Turning to *FBA*, Eq. (6.10b), one is maximizing the flux of cellular metabolic reactions as given by $c^T v$ where c^T stands for the data-based metabolic demand for biomass precursors and v stands for respective fluxes, the throughput, to be determined. The shown constraints are: (1) $Nv = 0$, the steady state (see Chapter 1: Change/differential equations, Eq. (1.19b)) with v as before and N being a stoichiometry matrix such as in Chapter 2, Consensus/linear algebra, Eq. (2.5), RHS. (2) $0 \leq v_i \leq \beta_i$, that is, each reaction flux, v_i, can take positive values up to a maximum, β_i. To carry out the optimization translates into solving the equation systems for their variables, x in Eq. (6.10a) and v in Eq. (6.10b), and identifying those that will maximize the overall output.

Mobilizing *LP* for optimization and using computers for the actual computations has practicalities of its own. These may be dealt with using mathematics that is useful also elsewhere and deserves here a comment. As it were, computing constraints in an equation system is often substantially more demanding than computing the variables. *LP* optimization problems often have many of the former and a few of the latter. Reversing this relationship for the sake of the computation would be beneficial, especially in large models. It is, in fact, feasible, turning to *duality*. As it were, certain mathematical expressions can be seen from more than one point of view, that is, represented by interconvertible alternative expressions. Such, in fact, have appeared earlier in this book—notably differential equations and their Laplace, Fourier and Legendre transforms as well as the expressions for observability and controllability in linear control theory. Turning to *LP*, as is the general convention, its equation system is said to be initially in its *primal* form. By appropriate manipulation it can be converted into its *dual*, the alternative point of view, when called for. The dual is naturally convertible back to its primal form which is the dual of the dual. An original, primal problem with, say, four variables and two constraints will thus become its dual problem with four variables and two constraints. Optimization of the dual form will yield values directly related to those optimal in the primal version. These may be interpretable directly or processed further into a solution corresponding to the primal problem.

How does *LP* identify optima? An answer may be available in a diagram such as Fig. 6.2, showing a hypothetical problem optimizing a performance measure as in Eqs. (6.9a)–(6.9c) with variables x_1 and x_2, but with four arbitrary constraints. Whether equalities by origin or after manipulation when not, there will be an area delimited by corresponding straight lines and the *x,y*-coordinates. The striated area in the figure delimited by the straight lines represents the set of solutions that are *feasible* while satisfying the constraints. Also known as the *region of feasibility*, the points O, A, B, C, D, and E are its *corners* (or vertices). As it were, optimal values, that is, optimal combinations of x_1 and x_2 will always be located at the *corners*—at an intersection of lines that represent constraints. This makes intuitive sense because the latter indeed represent extreme values and would be meaningless if the optimum is at an interior point. Simple *LP* cases can thus be solved graphically by locating the corner points and comparing the values of the corresponding performance indices.

At the computer end, solving *LP* models most likely will turn to algorithms based on Dantzig's *simplex* method and its numerous variants. The name of the method is due to the region of feasibility formally belonging to a mathematical entity known as a simplex, but of no further concern here. Applying the simplex algorithm will first generate a feasible solution at one of the corners then move recurrently to generate more corner solutions. Comparing each time with previous values, the process terminates when there is no further improvement to yield a global optimum.

Dynamic programming

As context for the second so-called programming approach, it will be convenient to consider first a practice that, offhand, is tangential to cellular process

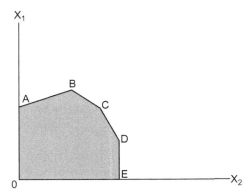

FIGURE 6.2 Corner optima.

behavior, but has a significant role in molecular cell research otherwise. Looking for sequence homologies between proteins or nucleic acids is a common way to identify unknown cellular components, functional similarities, or evolutionary relationships, and so forth. Sequence alignment is a process currently conducted by routine computer software that will evaluate how well sequences match. Largely unappreciated is that major alignment protocols are, in effect, quantitative process optimization procedures. Some with most basic mathematics, these methods embody yet another, quite different line of thought in the optimization of a wide range of real-world features. Taking up sequence alignment in this context here will thus expand on the infrastructure for processes and other optimization described above, also shedding light on a significant, but often obscure, aspect of its customary usage.

Extending on the various parameters chosen for optimization so far, sequence alignment procedures optimize information derived from amino acid or nucleotide residue similarity properties. Comparing residues of two sequences pairwise, one assigns a *quantitative* score to each comparison. What can be scored? In the most elementary case, see example below, one would score whether residues are identical or not, assigning the value one (1) when they are, and zero (0) when they are not. There are many possible refinements such as assigning values that differ depending on a perceived degree of similarity between residues, or penalizing for discrepancies due to extra or missing residues or larger sequences, among various others. Whichever, to determine similarity, scores for all residues will be summed up, those optimal defining the best alignment(s) between sequences, keeping in mind that as with much else in modeling, the optimum score depends on the choice of a model. Employing different similarity parameters could align different residues. Computer-implemented, sequence-homology searches can be demanding, as can be optimization procedures elsewhere. A common problem, especially with many entries, is that even when an optimum is found early on, it often remains hidden amid numerous other possible outcomes throughout much of a long routine.

Enter dynamic programming, essentially a strategy that allows to economize on computations by converting large optimization procedures into a number of smaller ones handled separately. Adaptable, it is already widely employed in technological and scientific applications such as optimizing resource allocation, developing industrial controls, modeling chemical reactions, and solving mathematical equations. Unlike LP, which is dominated by the simplex algorithm, dynamic-programming algorithms vary with the problem and can be deterministic—discrete or continuous, linear or nonlinear, or stochastic.

Originating in the mid-20th century, dynamic programming was conceived as a sequential *multistage decision process* in which a decision is taken specific to each stage, but so as to optimize an entire process.

Underlying is Bellman's *optimality principle* which can be applied in different ways. Here noteworthy is that when beginning an optimization procedure with a given initial stage, the principle stipulates that the remaining process will be optimal when the next stages are optimized consecutively, each on its own, but in a way that also reflects those previously found.

When applied to process optimization, dynamic programming could formally begin in a standard format with performance measures and constraints, as described above; but instead of aiming for optima for an entire lifetime of a process, as would, for example, solving Eqs. (6.6) or (6.8), optima would be computed as a sequence of *stages* that take place over shorter time intervals, that is, between points t_1, t_2, t_3, ..., t_n. Each stage is, again, initially in a particular *state*, defined by its *value*. To begin with, it has available a number of options for further progress with one or more possibly optimal. Each stage is thus essentially a subproblem with small scale computations to match, *embedded* in the overall optimization procedure. Altogether dynamic programming is sometimes represented by the scheme $f_1 \rightarrow f_2 \rightarrow \cdots \rightarrow f_i \rightarrow \cdots \rightarrow f_n$, where the fs stand for the function to be optimized in stages from f_1 through f_n. The arrow, $\rightarrow$, is a symbol for a mathematical relationship that is *injective*, a term that is here intuitively suggestive. It could, for example, represent values generated in a particular stage as being output injected as input into the next stage. For actual machine computation, dynamic programming procedures generally can be cast as recursion formulas. Note that the underlying logic of embedding partial optimization choices in a larger scheme appears to have counterparts also in actual cellular processes such as protein folding and others. Investigating whether Bellman's principle of optimality indeed has a wider role in cellular process design might add a significant component to our understanding of cellular systems.

How might the optimality principle work in practice? The *Needleman–Wunsch* sequence alignment algorithm [2] was conceived with computational demands in mind, making use of the same $f_1 \rightarrow f_2 \rightarrow \ldots \rightarrow f_n$ scheme. A prominent prototype, it has carried over to more recent alignment programs including popular searches such as fast-all (FASTA) and basic local alignment search tool (BLAST). In the example shown in Fig. 6.3, reproducing Fig. 6.2 of the original paper [3], the alignment compares residues of two amino acid sequences. The residues are represented by the common single-letter amino acid code with the less customary symbols B standing for D (aspartic acid) or N (asparagine), and J for L (leucine), or I (isoleucine). Juxtaposing the two sequences in a matrix, as seen, each possible pair, matching or not, has its own address in a particular row (horizontal, numbered from top to bottom) and column (vertical, numbered from left to right).

The actual comparison takes place in stages, one residue at a time, and scored. When residues match, the score equals one, when they do not it equals zero. Gaps between matching residues are ignored. Scoring begins at

	A	B	C	N	J	R	Q	C	L	C	R	P	M
A	8	7	6	6	5	4	4	3	3	2	1	0	0
J	7	7	6	6	6	4	4	3	3	2	1	0	0
C	6	6	7	6	5	4	4	4	3	3	1	0	0
J	6	6	6	5	6	4	4	3	3	2	1	0	0
N	5	5	5	6	5	4	4	3	3	2	1	0	0
R	4	4	4	4	4	5	4	3	3	2	2	0	0
C	3	3	4	3	3	3	3	4	3	3	1	0	0
K	3	3	3	3	3	3	3	3	3	2	1	0	0
C	2	2	3	2	2	2	2	3	2	3	1	0	0
R	2	1	1	1	1	2	1	1	1	1	2	0	0
B	1	2	1	1	1	1	1	1	1	1	1	0	0
P	0	0	0	0	0	0	0	0	0	0	0	1	0

FIGURE 6.3 Sequence alignment.

the bottom row and progresses upward, adding scores. At each consecutive amino-acid pair, the score represents the maximum number of matching pairs found so far. Thus the first comparison has a match at P, row 12 column 10, no matches at the next residue, row 11. Now R, row 10, has three matches at columns 1, 6, and 9, but only one can be a candidate given this is a sequence, so that the maximum of matches in that row is 2. Once scoring in this way is complete, the best sequence alignment will be determined starting at the maximal score, here at the top-left corner, row 1 column 1. As shown by the arrows, it will proceed, tracing the maximal possible overall score, here incidentally having two options for an optimum. The best alignment in Fig. 6.3 is thus at residues A−C−[J or N]−R−C−C−R−P.

Keep in mind

By quantitative measures, processes are *optimal* when they attain a *maximum* or *minimum* value of a property of interest. Being optimal or *optimality*, as known in cellular processes so far, generally falls within two prototypes. A best choice of a *single* route out of several concurrent processes that differ in their rewards. Otherwise, a most favorable *balance* between incompatible processes. *Optimization*, evaluating the most favorable conditions depends on the particular case. Altogether, evaluating optimality in cellular processes can take different routes. The same type of problem may be open to different approaches; optimizing cellular biomass production is a common candidate.

Formal optimization procedures in this chapter optimize a *performance measure/objective function* that can be based on process material properties or on other characteristics of practical or theoretical interest. *Constraints*, internal and external, often dominate real-world optimizations. Generally represented by separate equations that *augment* the performance measure, *equalities*, possibly using *Lagrange multipliers* are preferred while *inequality* constraints, significant in optimizing real-world processes, are more cumbersome and can become problematic.

Classical optimization is applicable to cellular processes when available information is sufficient to describe the performance measure by a unique mathematical function. To identify minima and maxima, they employ *derivatives* of a performance measure, as in differential calculus.

Variational methods, mainstays of formal optimization, can handle the more common real-world circumstance in which process information is initially limited. Performance measures represent a range of plausible options, a functional of *admissible* mathematical functions. The one optimal will have a maximum or minimum value of a definite *integral* of the chosen model. It will be identified based on conditions derived from the mathematics of the first *variation* of the performance measure, an analog in the first derivative similarly used.

In the *calculus of variations*, the original variational approach, the primary condition for being optimal is that function in the performance measure satisfies the *Euler–Lagrange* equation. Procedures typically apply to single or multiple, essentially independent, processes with well-behaved constraints. Extending conceptually as well as in range applications is the *Hamiltonian* format, originally a technological prototype for optimizing primary processes strongly interacting with their controls. The associated *Pontryagin maximum principle* stipulates that given an optimal primary process the control is required to be at its own optimum at all times for the duration of a process.

LP is tailored to optimizing constrained processes described or approximated by linear, algebraic equation systems, a format attainable from a variety of other types of process descriptions. Represented geometrically, possible solutions of the equation system are found in a *region of feasibility*, a polygon bounded by the coordinates and curves of the constraints. Those optimal are located at *corners* where curves of the polygon intersect. Transforming the original, *primal* equations into a *dual* form can offer significant computational advantages. Using computers, optima are identified with *simplex*-type algorithms.

Dynamic programming is a wide-ranging strategy for handling large problems with extensive demands on computer power. It allows to economize computations by breaking large optimization procedures down into sequences of smaller subprocedures. *Embedded* in the overall scheme, these smaller scale *stages*, are computed separately and sequentially, also taking into account previous stage histories. According to Bellman's *optimum principle*, optimizing each stage on its own, in turn, will add up to an overall optimum.

References

[1] Cohen D, Parnas H. An optimal policy for the metabolism of storage materials in unicellular algae. J Theor Biol 1976;56:1–18.

[2] Gopal M. Modern control system theory. 2nd ed. New Delhi: Wiley Eastern Limited; 1993. New Delhi: New Age International Publishers, Reprint 2010. Chapter 10.

[3] Needleman SB, Wunsch CD. A general method applicable to the search for similarities in the amino acid sequence of two proteins. J Mol Biol 1970;48:443–53.

Further reading

Krylov VI. The calculus of variations. In: Aleksandrov AD, Kolmogorov AN, Lavrent'ev MA, editors. Mathematics: its content, methods, and meaning. Moscow: Russian Academy of Science Press; 1956. English: Cambridge: MIT Press, 1963, Mineola: Dover, 1999, Chapter VIII.

Pierre DA. Optimization theory with applications. New York: John Wiley and Son; 1969. Mineola: Dover, 1986.

Pinch ER. Optimal control and the calculus of variations. Oxford: Oxford University Press; 1993.

Chapter 7

Chance encounters/random processes

Context

Inconvenient truths? Rogue behavior? Cell biology has its share. Routine processes in cellular life such as those discussed so far in this book are taken to be deterministic, continuous, and well behaved. Although not unheard of before, recent technological advances have increasingly uncovered biologically significant cellular processes that do move forward, but appear to be *random* or *stochastic*, discontinuous, and more or less erratic. Not what you expect of complex, but highly integrated, coordinated, and presumably efficient systems. Certainly not in biological cells if randomness can also lead to aberrant behavior and pathological consequences. To accommodate a relatively recent paradigm in cellular process behavior and its modeling, a short detour from the otherwise deterministic outlook of this book will in this chapter turn to stochastic process descriptions and associated infrastructure.

What are typical indications of random behavior? Activities that seem uncoordinated, data points that are scattered beyond any measurement error and difficult to reproduce, knowing one point does not tell with certainty where others may be. Real-world random processes typically fluctuate around an average trajectory, instability that can be problematic in technological processes, as has been seen in previous chapters. Altogether, stochastic behavior in natural and artificial complex systems is often due to random environmental effects. Internal factors—imperfections in the design and workings of complex systems—have a role as well. Randomness is, however, also physically inherent in molecular level behavior, as seen in small numbers (see Chapter 1: Change/differential equations, footnote 3).

Major cellular processes take place at the molecular level, and, as readers well know, are indeed fundamentally stochastic. In large numbers, as commonly studied, they display narrow statistical averages and are well behaved. At the same time there are relatively *rare* special events in cellular biology that for some time have been considered random—for example, spontaneous and chemical mutagenesis, DNA recombination used as a measure of genetic distance, and the biological actions of radiation. There is increasing evidence for random fluctuations in the numbers and activities of cellular components

Collective Behavior in Systems Biology. DOI: https://doi.org/10.1016/B978-0-12-817128-8.00007-9

that are routine. Call it *noise*, it is being implicated in routine behavior, in phenotypic differences between isogenic cells, as well as in more specialized functions such as cell differentiation and latent virus activation. Random behavior is especially well documented in single-cell DNA transcription in pro and eukaryotic cells. While external random system fluctuations have a role in these processes, intrinsic properties of the participants appear to be significant. Also pointing in that direction are random variabilities elsewhere, in catalysis by single enzyme molecules and in chemoattractant—receptor interactions. More of the same behavior could arguably emerge as experimental methods develop further. On the other hand, cells appear to have also mechanisms to suppress noise. How common and what significance random behavior has in the everyday cell biology and in evolution or in various pathologies, is still very much an open question.

Now, though this be randomness, there is method in it. Presumably depending on *chance*, random processes can still be understood and modeled, albeit in *probabilistic* or *statistical* terms with infrastructure to match. It may, to begin with, be possible in essentially the same or related formats used with common, macroscopic, models. Minor random input into natural and artificial processes, *noise*, is occasionally represented by adding special *correction* terms, common practice in modeling technological processes with differential equations. When both deterministic and stochastic contributions are significant, *stochastic differential equations* (see Chapter 1: Change/differential equations) may be favored. Common in *molecular dynamics* simulations of change in proteins and nucleic acids, a powerful, but resource intensive modeling approach, are those of the *Langevin* type. Recall also that *chaotic* systems with their extremely random appearance are nevertheless due to the combined actions of deterministic factors, nonlinear that they are, and representable as such.

Models of random cellular processes discussed here will be probabilistic throughout. Quite elementary probability notions, briefly reviewed first for those interested, sometimes go a long way. Here they will allow to present a prototype that can describe major processes in nature and the artificial world, the *Poisson process* model. This will be followed by an increasingly popular tool for simulating low-copy cellular and chemical processes that builds on that prototype, the Gillespie *stochastic kinetics simulation algorithm*. As described here the Poisson model and the Gillespie algorithm apply when processes' participants are completely independent and free to be or not to be active. Probabilistic process models can also take into account system members that interact, curtailing their independence, a situation more common in complex systems. An example will introduce another wide-ranging approach to statistical modeling. Centering on *partition functions*, it will be applied here to a biologically significant prototype, the *helix—coil* conformational transition in polypeptides. Keeping the abundant detail to a minimum in both examples the focus here will be on the motivation behind.

Low copy number processes

A single cell setting

Cells communicate and to some degree coordinate activities. In eukaryotic tissue obviously, but also when living in a colony and most likely those existing as solitary microorganisms as well. Processes that behave randomly are expected of cells or functional units that, on the other hand, are independent and present in small numbers. What are the relevant cellular statistics? In eukaryotic cells, components are often present in the thousands or several hundreds. In prokaryotes, however, the copy number of certain enzymes or m-RNA species or even genes can be down into single digits. Among cellular components, generally at the lower end, are the machineries of transcription initiation, intracellular signaling, and transmembrane ion transport. Some of these can be monitored as single, functional units.

As a rough guide, random variation in the activities of n functional units is often taken to occur in proportion to $1/\sqrt{n}$. For $n = 10,000$, incidentally an order of magnitude estimate of higher-end enzyme copy number per cell, this translates into a 1% deviation. With 100 and 10 members it will be 10% and 33%, respectively. More precise estimates of random deviations (*error*) in real life vary with system make up, size, and complexity and could exceed this range significantly. Given these numbers, while deterministic models generally represent order of magnitude higher participant numbers—10^{10} in a test tube is not unusual—low copy single cell processes naturally fall within so called microscopic behavior (see Chapter 1: Change/differential equations, footnote 3), commonly described in statistical terms. The Gillespie algorithm is a relatively recent prototype for modeling processes as random reactions that take place in small numbers, one reactant (molecule, atom, ion) at a time. Leaving alone the substantial computational detail, what follows will lead to the statistical process description it implements. Readers wishing to refresh on probability notions this can take will find the next section helpful.

Probability

Basic notions

A major concept in statistical process modeling is already familiar, namely that of a *state* with its attached values of process parameters. As with the deterministic *state variable* model of technological systems (see Chapter 4: Systems/input into output) a stochastic process can be viewed as an initial state that is followed by a sequence of momentary states, but now given in terms of probabilities.

In simplest terms *probability* is a measure of the *chance* of a random system being in a particular state. The latter is known as an *event*. Although not accounting for all the options available in the real world, probabilities in nature and artificial systems are often presented as a choice between only two options, that is, a *yes* or *no* answer to whether or not an event actually took place. It is mathematically convenient and will here and in Chapter 8, Organized behavior/networks, go a long way. Thus whether or not, say, DNA synthesis in dividing cells was observed technically corresponds to *success* or *failure*. With p and q as the probabilities of "success" and "failure," $p + q = 1$. The total probability of success (yes) can then be taken to equal one while total failure (no) will equal zero. In terms of actual numbers, the probability of success equals the *frequency* of the particular event—the ratio of its occurrence relative to the occurrence of all related events. Given, say, a cellular life cycle of 24 hours and a DNA synthesizing s-phase of 8 hours, the random probability of observing DNA synthesis is $8:24 = 1:3$. The probability of failing to observe DNA synthesis is $16:24$, or $2:3$.

In the real world there is frequently a need to evaluate *joint* probabilities, say, of two or more consecutive processes or, for that matter, the joint occurrence of more permanent structural features. Consider incorporating RNA polymerase into a transcription preinitiation complex, call it process X, and the subsequent binding of the complex to a DNA promoter, process Y. The probability of observing the association of the polymerase with DNA will be given by the *joint* probability of the two processes. Using the formula $Pr(X,Y) = Pr(X) \times Pr(Y)$ it is also known as the *multiplicative* law of probabilities.

Information on a particular component of a process may become available during a study or may have existed already before elsewhere. It could simplify evaluating that which remains unknown. Taking the transcription example one step further, suppose that one knows already that the polymerase is present in the complex, what are the chances of finding it also bound to DNA? Known as *conditional* probability $P(Y|X)$, the probability of Y given X, is defined as $P(Y,X)/P(X)$. A formal way to express dependencies between states, it is an important tool in modeling random processes.

Chances of individual members in a population

Reckon that the above probabilities of success and failure in observing the cellular s-phase, for example, are properties of a cell population, usually in substantial numbers. Individual cells, however, have no such choice; they only can be either a success or a failure. Given a population of cells with a life cycle such as the above, what is the chance of a particular cell being in the s-phase? A molecule of a particular substance reacting? How many in a small number cell sample will be synthesizing DNA?

Doing the computation naturally takes into account both the probability of success and that of failure in the entire population. However, calculating probabilities, and this is a recurring feature, also assumes that when there is less than one success per system member, technically $P < 1$, success is equally likely in all members of a population or a sample, and one has to consider also in how many ways this could happen. For example, when probability of success is one in three, one success can appear in any one of three cells (i.e., in three ways).

When, as in the above cell cycle case, the probability of success and failure are *comparable*, the probability of success among single cells in the population is likely to be given by the ubiquitous *binomial* formula. Here relevant is that in many real world cases, notably chemical reactions, the model process for the Gillespie algorithm can be modeled as *low* probability (of success) events. In other words, success occurs only in small numbers out of large populations of potential reactants. These models are based on the *Poisson approximation* to the binomial formula which can be used in different ways. For example, inactivating bacteriophage with UV light, computing the probability of exactly k lethal hits (success) using the Poisson formula can be given by Eq. (7.1).

$$p(k) = e^{-\lambda}\left(\frac{\lambda^k}{k!}\right) \tag{7.1}$$

On the LHS of Eq. (7.1) is the probability, P, of k lethal hits. As given on the RHS, it depends on the parameter Lambda, λ, a measure of the fraction of the total radiation that is actually lethal. Lambda is here essentially the low probability counterpart of the above total number of cells and the probability of success, say, observing DNA synthesis. Also on the RHS, e is the basis of the natural logarithms, 2.718, ..., and $k!$, which denotes k *factorial*, is the product of all integers from 1 through k, that is, $k! = 1 \times 2 \times 3 \cdots \times k$.

Population profile

Information on stochastic process behavior is available also from a collective profile of specific event probabilities. Taken together and often represented graphically, these make for a *probability distribution*, a profile of say the probabilities of 0, 1, 2, 3, 4, and so forth, successes in delivering phage lethal hits. Not surprisingly Eq. (7.1) is the formula for the *Poisson distribution*. The Poisson and parent binomial distributions are discrete—the success values differ by whole integers. Continuous probability distributions are common as well; the familiar, bell-shaped normal or Gaussian distribution for example.

Journey stochastic

Proceeding to describe stochastic processes quantitatively it will be helpful to take another look first at what was typical in the deterministic process models of previous chapters. In these models, transitions between states (i.e., change both in process *values* and *timings*) were generally regular, gradual, and predictable. This was so whether the process was continuous or discrete.

The defining feature of the model was the process value, one for each point in time. Moreover, the *initial* state was the same for all participants and change occurred in one *direction*, commonly forward.

Stochastic models extend on these options and can be flexible about what they combine. In some types of process, values may be multiple and taken at random; in others change is single valued constant. Time intervals may vary randomly or be regular. Ditto initial values, direction, and possibly other features. Variables in the deterministic models are usually in the same format while random process models may be mixed—continuous in one feature, but discrete in another. As with deterministic models, probabilistic rate laws may stay constant during a process or they may change.

That a real-world process on its way may also depend on previous history of the system is, offhand, conceivable. An issue of having *memory* of past states, it could be ignored modeling deterministic systems so far (see Chapter 4: Systems/input into output), but often needs to be clarified in modeling stochastic processes. It turns out that many a real-world random process behaves as if activities leading to the next state of the system depend only on its current status, but not on any state before. Thus an enzyme using, say, glucose as substrate minds only those glucose molecules currently in its presence, but is oblivious to glucose that was present before, or for that matter by what route it came about. Recall that the same holds for the deterministic behavior of large-scale processes such as chemical reactions, spontaneous radioactive decay, or log phase microbial cell division, among various others. Altogether known as *Markov Processes* they are considered to be *memoryless*. Technically, if the system at time n is at state y_n, the conditional probability of moving to state y_{n+1} is independent of whether state y_n arose directly from the given initial state, y_0, from the previous state, y_{n-1}, or from an earlier intermediate state, y_{n-k}. As with deterministic descriptions, this is another modeling that is desirable.

Fortunately for modeling, major types of natural and artificial random processes can be taken to occur in only a small minority of system members and have their natural match in the convenient mathematics of the *Poisson process* prototype. Extending on the Poisson formula, Eq. (7.1), a process defined by the probability of time-dependent state transitions can be given by Eq. (7.2a).

$$p(k_t = k) = \frac{e^{-\lambda t}(\lambda t)^k}{k!} \tag{7.2a}$$

$$p_\tau(t) = \lambda e^{-\lambda t} \tag{7.2b}$$

Using the same symbols as Eq. (7.1), which it obviously resembles, Eq. (7.2a) specifies the chance of k successes happening during time, t, here the time that passed from the beginning when $t = 0$. λ in Eq. (7.2a) is a rate parameter that equals the average number of state transitions known to take place over an extended time period. Change in process values, also known as *jumps*, is *discrete*, predictably *one* unit at a time throughout.

Now, physical time, that experienced by the real world, obviously remains continuous, moving forward regularly. Continuous models allow state transitions to occur anytime and last for an unspecified duration. The time the Poisson process model spends in different states, however, is predetermined and varies. Time flow in the Poisson model is thus discontinuous and time adds up to a sequence of probability dependent intervals. The length of each interval, among others called *sojourn* time, is random, obeying the *exponential* distribution, Eq. (7.2b). Similar timing schemes are common in modeling random, Poisson type processes also elsewhere (see Chapter 8: Organized behavior/networks).

Gillespie's simulation model

A modeling tool that is based on the Poisson process and is highly mindful of the computational aspect is the Gillespie *stochastic kinetics simulation* algorithm [1]. The Gillespie algorithm has, among others, been applied to model cellular processes, notably enzyme reactions, and the various phases of gene expression and associated detail. First developed to describe chemical reactions, the Gillespie algorithm originated recognizing that conventional models, deterministic, often continuous, geared toward large number averages can be deficient because in reality single reactions would be random and discrete. Both on general principles as well as to capture detail such as natural fluctuations hidden in averaged large populations, process models should therefore be microscopic, discrete, and probabilistic. As the name indicates, however, this is not a model that will be solved analytically, but will be worked up as a numerical simulation in the sense of Chapter 1, Change/differential equations, and Chapter 3, Alternative infrastructure/series, numerical methods. Based on detailed physical, mathematical, and computer-related considerations there are by now a number of versions of the Gillespie algorithm.

As do common process models the Gillespie algorithm computes the progress of a chemical reaction as change in the amounts of participants over time. Change here depends on reaction probabilities, the chance of a transition in a chemical state. Recall that probabilities have an already implicit role in the familiar textbook view of chemical reactions. Expressed by mass action and rate laws, models reflect an average large population probability of particles (molecules, atoms, ions) reacting over continuous time. The microscopic approach of the Gillespie algorithm, on the other hand, is concerned not with averages but with the explicit probabilities of what happens when single particles react. In the typical case of two different reactants, chance is expressed by the probability at a given time of having available in a large population—say of two proteins forming a complex—a small number of encounters between appropriate participants. The actual number of encounters could vary, as given by a probability distribution. That the encounter is actually productive once it happened, say a stable complex is formed, happens only in some of the cases.

Turning to the time factor, while the passage of physical time is again taken to be continuous, the actual timings of each productive encounter are discrete, unequal in length, and taken to be random as given by the exponential distribution. For the procedure to be physically and mathematically meaningful, both the participants and the time intervals are chosen randomly using ancillary procedures. Altogether, as with conventional models, the Gillespie approach takes a chemical reaction to be a Markov process (see the section: Journey stochastic, above). This calls for specifying the current status of the process as well as the properties of the chemical transition as just described.

At the quantitative end the process is ultimately represented in the algorithm by the two above factors. First, the (instantaneous) probability of a chemical reaction taking place, $P(\mu)$, will be given by Eq. (7.3a). As in Eqs. (7.2a,b) the probability of time behavior, $P(\tau)$, the second, will be provided by Eq. (7.3b). Using the multiplicative law, the joint probability, $P(\tau, \mu)$, is the desired centerpiece of this approach, taking the form of Eq. (7.3c).

$$P(\mu)d\tau = a_\mu = h_\mu c_\mu \tag{7.3a}$$

$$P(\tau)d\tau = exp(-a_0\tau) \tag{7.3b}$$

$$P(\tau, \mu) = a_\mu exp(-a_0\tau) \tag{7.3c}$$

In Eqs. (7.3a,b,c) on the LHS the variable μ represents the probability of a productive reactant combination, the variable τ represents the time. On the RHS, beginning with Eq. (7.3a), the first term, a_μ, is the probability of the reaction taking place. It is a rate parameter, analogous to λ in Eqs. (7.2a,b), here combining two parameters. h_μ represents the sum of distinct productive combinations among candidate reactants only, say effector or receptor pairs, possible among the total present, and c_μ represents a reactivity parameter based on specific properties of the reaction. Parameter a_0, for reasons detailed in the original, is calculated as the probability of the whole complement of reactants *not* having reacted in the stage prior to the reaction taking place, assumed in the model to be the great majority. It is similar to a_μ except that it stands for *all* possible reactant combinations even if unproductive.

Using for the actual computations the joint probability, Eq. (7.3c), further handling is by dedicated computer routines. Given a number of starting reactants, these employ parameters described above to calculate the probabilities of the occurrence of single reactions with time, choosing candidates and timings by randomizing *Monte—Carlo* subroutines. After each single reaction the number of reactants remaining and the time of the next reaction are recalculated, updated and used for the subsequent reaction, and so forth.

A run iterating this protocol describes the course of a reaction. Repeating the same run will usually generate trajectories that differ in some detail. A sufficient number of reiterations should provide a good average value and an estimate of the extent of natural fluctuations of the system. The Gillespie algorithm is also applied at higher cellular levels, for example, in assessing process variability between single cells in a population.

Interacting neighbors

A polypeptides environment

Poisson processes in systems such as where reactants are essentially on their own, fit nicely into what is expected of random interactions in a process and can be described in elementary terms. When reactant behavior is subject to additional, more specific interactions, stochastic modeling becomes less straightforward. It is still possible, nevertheless, among others using models that are both biologically relevant as well as belong to a more general approach in statistical modeling well worth a comment.

Polypeptides, small oligopeptides to large proteins as readers will recall, can assume a variety of conformations. These include structures that are broadly classified as helices, beta-sheets, and random-coils. A particular polypeptide can be completely in one of these states or host a mixture, some or all of the time. The different states and their transitions, in turn, reflect differences in the conformation of their amino acid building blocks. Residue conformations also determine which and how different building blocks will interact. Which particular conformation a certain amino acid residue or sequence of residues adopts can depend, among other factors, on the temperature, the solvent, and various solutes. When it comes to quantitative modeling, conventional macroscopic models would be a natural first choice. Deterministic, models unexpectedly happen to fail, however. Statistical alternatives fare better, the basic idea being to feature the *probabilities* of neighbor interactions affecting conformational properties of the amino acid residues and the polypeptide.

Selecting next neighbors

A favorite model system in these studies is the polypeptide *helix−coil* transition. The model polypeptide can be inhabited by stretches of various lengths of either *h(elix)* or *c(oil)* conformation, offhand distributed randomly over the entire length of the polypeptide chain. As an arbitrary example, consider *cccchhcchhhch*. Physically intuitive and a major player in modeling helix−coil transitions is the *Lifson−Roig* (*LR*) model [2]. Preceded by other polypeptide models, it ultimately originates in the *Ising* model of neighbor effects on the magnetization of arrays of magnetic elements.

The *LR* helix−coil transition model naturally makes use of the conformational properties of the amino acid residues, but also considers properties of their close neighbors in a peptide and how they might interact. Underlying is a putative scenario of how a polypeptide such as that above would have been constructed in a sequential process taking into account certain rules. As can be seen earlier, there is to begin with a stretch of at least three amino acid residues in the coil conformation, favored. Assigning a conformation to the next residue, here say residue 4, it is highly likely that it will also assume the coil conformation. The probability is low that the next amino acid, here residue 5, is the *first* in a helical conformation, energetically a difficult *nucleation* step. Assuming for the sake of the argument that nucleation, nevertheless, has taken place, placing the *next* residue (6) in a helical conformation, *propagation*, should be easier and is given a higher probability. With three consecutive helical residues, for example residue 11, a *hydrogen bond* will be formed across residues that is not present in the coil conformation. As it were, a change from one type of conformation to the other is also disfavored in the opposite direction and the probability of a helical residue followed by a coil residue, residue 13, is again low. Considerations such as these would also affect transitions, for example, once the temperature changes.

What can a system such as that reveal about polypeptide behavior? Obviously, models should mimic experimental observations. Validating the model and its practical applications, this will also enforce the notion of neighbor effects, and perhaps, as does Gillespie kinetics, also make a more general point for statistical modeling. As it were, the *LR* and other models, provide standard system parameters that would be of interest regardless of neighbor effects such as average lengths of helical regions and the distribution of their sizes. In some cases, one may discriminate whether averaged values represent mixtures of molecules entirely in one conformation or such where helical and coiled regions appear simultaneously in the same molecules. The degree of intrapolypeptide hydrogen bonding, the ratio of hydrogen bonds to total residue number, is another parameter of interest.

Energies into probabilities

Now, aiming to use the model and actually compute polypeptide properties from the probabilities of the various amino acid residue conformations and the associated neighbor effects, one obviously would like to know where these probabilities would come from in the first place. Here enters the notion originating in statistical physics, a postulate that the probability of member arrangements in a statistical system are tied to their energy. How might energy and probability be related? In the most elementary terms and in the simplest system of an ideal gas, particles are taken to behave independently, randomly, and likely to occupy various energy levels, E_i, where i is an index. The probability of a particle occupying energy state E_i is given by $e^{-Ei/kT}$

where e is the natural logarithm base, E_i is the share of the energy state, i, in the total energy, E, strictly speaking of a most probable of possible arrangements, k is the Boltzmann constant, and T is the absolute temperature. The sum total of all particle probabilities is its *partition function,* $Z = \sum e^{-Ei/kT}$ where $\sum$ is the summation symbol. Underlying the partition function is thus an energy-dependent probability profile of the entire system. As readers may already know, partition functions are a versatile tool for calculating overall system thermodynamic properties from those of the lower-level constituents. They take various forms depending on the system and its model as is the case here.

The source of helix−coil conformational probabilities, the energy taken by the *LR* model to represent conformational behavior, is the *configurational* energy that is associated with bond rotation around the angles of rotation of two amino acid side chain bond angles. Available in the literature, these parameters differ between the helix and coil conformations and affect the probability of residues being in either. Unlike the discrete gas particles above which occupy distinct energy levels and whose partition functions represent a sum of probabilities, amino acids can assume a continuum of rotation angles and, thus, of associated energies and probable configurations, leading to partition functions given by continuous *integrals* of similar expressions. Amino acid residue partition functions are integrals of exponential functions in the $e^{-Ei/kT}$ mold where E_i represents, configurational, hydrogen bonding, and solvation energies as the case may be. Partition functions for entire poly-peptides, combining the configurational energies of the amino acid residues can be calculated as well.

Polypeptides harboring helix−coil transitions are described in the *LR* model in similar terms, taking into account that amino acid residues are not completely independent, but are restricted to a chain in a polypeptide and have close, interacting neighbors. The total configurational energy of the system still remains the sum over all residues. The partition function, however, represents not independent probabilities for each residue conformation, helical or coil, but probabilities of each residue *conditional* on certain rules. The *LR* model thus assigns the probability u to a coil state, v, to coil or helix residues followed by the other conformation, and w to a residue interior to a helical stretch. Overall the partition function is a *product* of all the possible joint contributions of each of the residues. Turning once again to the peptide *cccchhcchhhhch*, the overall partition function would be the product of the probabilities *uuuuvvuuvwwvuv*.

Solving the polypeptide partition function of the *LR* model, generally elaborate, provides the statistics above and also allows to characterize the effects on conformational behavior of polypeptide properties such as chain length or the relative magnitudes of the various contributions. Especially suited to work with single amino acid homopolypeptides, for example polyalanine, statistical models can describe helix−coil transitions in heterogeneous polypeptides and polynucleotides as well.

Keep in mind

Biologically significant, low copy number cellular processes are under increasing scrutiny. Exhibiting random behavior and calling for *probabilistic* representation, they find a match in the popular *Poisson* process model in which small numbers out of large populations of independent reactants change state randomly and *memoryless/Markov*. The *Gillespie stochastic kinetics simulation algorithm*, developed to compute chemical reactions, simulates a Poisson process by representing the probability of a small number of reactions taking place as the chance of the sum of possible distinct reactant productive combinations, considering also physical reactivity parameters specific to the reaction. Reaction timings are given by the *exponential distribution*. State transition probabilities and timings are chosen by randomizing *Monte—Carlo* protocols. Repeated execution of the procedure provides a measure of its random fluctuations. It is also applicable in assessing process variability between single cells in a population.

Random behavior modified by *specific* interactions between system members is observed in polypeptide *helix—coil* transitions. Assuming sequential addition of amino acids residues and based on bond rotation, *configurational* energies, the *Lifson—Roig* model assigns a high probability to a coil residue conformation being followed by the same, *propagation*, a low probability to coil—helix and helix—coil sequences, *nucleation*, and intermediate values to helix—helix sequences. Creating polypeptide *partition functions* that link residue and polypeptide configurational energy to the probabilities of helix—coil conformations opens the way to calculating various molecular parameters characteristic of polypeptides in solution. Polynucleotides are open to similar types of analysis.

References

[1] Gillespie DT. Exact stochastic simulation of coupled chemical reactions. J Phys Chem 1977;81(25):2340—61. Available from: https://doi.org/10.1021/j100540a008.

[2] Lifson S, Roig A. On the theory of helix-coil transitions in polypeptides. J Chem Phys 1961;14:1963—74. Available from: https://doi.org/10.1063/1.1731802.

Further reading

Elowitz M. Laboratory, <www.elowitz.caltech.edu/publications.html>.

Feller W. Introduction to probability theory and its application. 3rd ed. New York: J. Wiley & Sons; 1967.

Kolmogorov AN. The theory of probability. In: Aleksandrov AD, Kolmogorov AN, Lavrent'ev MA, eds. Mathematics: its content, methods, and meaning. Moscow: Russian Academy of Science Press; 1956. English: Cambridge: MIT Press; 1963, Mineola: Dover; 1999, Chapter XI.

Chapter 8

Organized behavior/networks

Context

Where there is no vision the people shall go awry (Proverbs 29:18). Biological cells indeed function by following programs. Being genetic, these programs specify and coordinate the collective behavior of cellular components—which will be active, how much, and when. It is hard to imagine collective behavior in the cell cycle, cell differentiation, or apoptosis without their guiding programs, let alone energy capture by mitochondria and chloroplasts, gene expression, metabolic pathways, or enzyme catalysis. Built into the genetic programs are the ways in which cellular functional units and their processes *connect* and *interact*—in other words, their *functional relationships*. The cellular components and their relationships taken together make for an overall scheme, an *organizational structure* (*architecture*). Presumably underlying is a *design* that is guided by task oriented *logic*. The role of organizational structure in cellular process modeling is the second main theme of this book. Increasingly employed in systems biology and encountered on occasion in previous chapters, modeling organizational structure will be taken up here in more general and systematic ways. Input from infrastructure already available elsewhere, mainly technological, will have a substantial role as before.

What qualifies in a cell as a functional relationship? Recognizable at all levels, there could be two or more proteins that bind, enzymes that are linked through a common pathway intermediate, or genes occupying the same chromosome. Functionally related as well are mitochondria supplying ATP to the cytoplasm, cell-cycle control apparatus and cellular phase transitions, cellular growth and storage machineries that compete for the same nutrients, antibody−antigen interactions.

Functional relationships vary in nature. Bound proteins, say, the subunits of hemoglobin, associated as they are, can be seen as connected by a *static* interaction, sometimes referred to as *passive*. On the other hand, enzymes linked through a common intermediate or mitochondria supplying the cytoplasm with ATP are connected by an ongoing process, a *dynamic* or *active* interaction. Also included can be interactions that can *potentially* take place but are not realized as yet: between two complementary, but separately

Collective Behavior in Systems Biology. DOI: https://doi.org/10.1016/B978-0-12-817128-8.00008-0

located nucleic acid sequences, foreign antigens and innate immune systems, sperm and egg cells, and, for that matter, a key and a lock.

What role could organizational features have in modeling cellular processes? The answer will take a bit of a journey beginning with model processes described in previous chapters. Reckon that these were intentionally simple or simplified for instructive purposes. Example processes were drawn mostly from the activities of basic cellular components and their reaction pathways. Even complex processes such as glycolysis in intact cells could be assumed to take place, allowing to ignore modeling problematic factors such as the particular sequence in which pathway reactions occur or intracellular barriers to the free movement of cellular components. In any event, these processes were taken to move predominantly along a *single path* in one, overall, forward direction. They would begin and end at the same time.

Intact cells and higher level subcellular systems, however, present an entire *landscape*. They harbor multiple and diverse processes that move in various directions. Pathways *intersect* and *branch off* at common intermediates—especially well documented in metabolism but known also elsewhere. Conversely, multiple cellular pathways *converge*—notably in signaling that regulates gene expression. *Crosstalk* takes place between signaling cascades and *feedback* is common from the enzyme level to the balancing of cellular growth and replication. *Barriers* to interaction posed by cellular structures may be significant. Unlike processes in previous models, *timings* are a major factor—diverse major cellular processes take place simultaneously, at comparable rates, and often starting and stopping at different times. Simplifying assumptions of previous chapters that allow ignoring the sequence of cellular events are not justified anymore. Cellular connection patterns depend on the particular case and can vary greatly. With parallels in the real world—the circuits of a computer chip, an urban water-supply system, the world wide web—*network* is a metaphor that captures well the organizational character of the more complex cellular systems. The notion of network, in general usage, applies also to considerably smaller systems and is a matter of structure and the types of processes that can take place.

Still, processes in networks retain system characteristics familiar from previous chapters. Whether they mediate material change or information transfer, networks use input, there is a transforming mechanism and output is likely. Processes are in a transient or steady state, possibly tending towards an attractor or oscillating, perhaps eventually approaching a halt. Familiar system issues also arise in network collective behavior—overall performance, stability to internal and external perturbations, susceptibility to control and its forms, adaptability to a changing environment, optimization. Altogether, present notions of system and network are naturally related and often overlapping; the terms are routinely interchanged in the literature when convenient, and occasionally here, as well.

The network environment, especially large and complex networks, can raise questions of its own. Much in modeling cellular network dynamics aims to reproduce process behavior in the first place with specific applications beginning to follow. It will be instructive to list typical questions, some already entertained in cell biology, that come from experience and theory in other disciplines. Questions can relate directly to the organizational structure and the membership: Is a network supporting one process or several concurrently? Is activity distributed over the entire network or is it localized to particular groups or single members and where? Is a network process dominated by a particular member, or a group of members? Which members interact directly and when? Do partial processes follow a specific sequence? Are particular biological functions associated with particular network members or types of organizational/network structure? What is the logic? Does a network have the right components and connections for a particularly desirable process to take place? Other questions arise about the actual workings of a network. Would concurrent network processes be cooperative? Compensatory? Competitive? Antagonistic? How do timings of partial processes affect the overall behavior? Will timings tend to a steady state? Would a process respond to interventions with the entire or with parts of a network? What will make a network reliable? Once perturbed will it recover? What will it take to disrupt its activity completely? Given that a process can take different routes, would all be useful? Efficient? Can the network be optimized? There are also questions of a more general nature. Would a process be sensitive to changing its parameters or might it be dominated by the network's organizational structure? Does a network with normal dynamics have a real chance to become chaotic? Under what conditions?

Models and infrastructure for network processes depend on the nature of the particular process, the purpose of the model, and on information that may already be available. There is no single, all-purpose format for describing these. A step up in complexity, the first choice is often still the kinetic modeling infrastructure common with simpler, forward moving processes, mainly single and systems of *ordinary differential equations* (*ODEs*), and their discrete counterparts. Considered the gold standard in modeling cellular dynamics so far, kinetic models are employed with networks whenever possible. They can be workable when design is not an issue, say, throughput in relatively small and simple networks or even in whole cells when modeling processes that dominate a network such as glycolysis in erythrocytes or various pathways in yeast carbohydrate metabolism. Differential equations also have a long history in modeling electric networks.

An obvious incentive to employ organizational structure in cellular process modeling explicitly would be to explore its role in cell biology, an area still in its beginnings. Networks could be small—controlling gene transcription—and extend up to control of cellular life cycles and differentiation. Modeling organizational structure is a practical necessity where kinetic modeling is inadequate, notably processes in large-scale networks with their complex interactions and diverse timings. Descriptions of such processes include the comings and goings in whole cells of ATP or various coenzymes, as well as major events such as mitosis or cell death. Employing organizational structure can provide an overview of process behavior and can facilitate dealing with the more local detail of network behavior.

Using organizational features is sometimes an only option for any kind of process modeling. Models such as these, *qualitative*, disregard the passage of time, but still can reproduce key qualitative features of cellular network processes and provide information on specific collective network issues such as whether processes are feasible or reliable. Qualitative modeling that uses organizational structure is a natural choice for modeling cellular networks whose dynamics are relatively insensitive to varying the kinetic parameters being presumably dominated by organizational features. A common incentive for its use is that data on time behavior simply are not available. When these are available, some organizational structure models provide scaffolds for attaching time-dependent and other quantitative parameters to create full-fledged models.

A word is due the remaining major factor in network operation as well as modeling—time. In previous chapter models the activities of all process participants started and ended at the same time. This makes time a convenient, single independent variable, notwithstanding the fact that participant timings in networks vary. Fitting network initial models with various timings, previously described, is generally mathematically impractical. It also would have no place here in models in which time has no role. Instead, diverse network timings are commonly specified in the computer programs used with the model. Timing in some cases is tied to the process itself, say, it will move on when it reaches a certain state. In others, timings are generated independently by a special computer program, for example as they were in the Gillespie algorithm (see Chapter 7: Chance encounters/random processes).

Organizational structure models here will center on system members and functional relationships. The infrastructure for this type of modeling is rooted in *graphs*. A term used in various ways, here graphs will be representations by custom pictorial elements, as the name suggests. They will be created, handled and interpreted as specified by *graph theory*. Providing definite rules, it combines input from the human visual experience with mathematics from a wide range of disciplines, including those described in this book. Graph theory is applicable to modeling a wide range of static and dynamic real-world features, including networks at various cellular levels, an example encountered in Chapter 4, Systems/input into output. Mason's rule for identifying critical steps in enzyme action combines signal flow graphs with linear algebra. Especially noteworthy here is that graphs also offer *platforms* for other modeling approaches. In these approaches, alternative representations take over by extending the scope of cellular modeling, guided by rules of their own. These, nevertheless, remain more or less linked to graph theory in their concepts, language, and choice of devices. To introduce common conventions, terminology, and a general flavor of the methodology, this chapter will begin with graph basics that can be useful in organizational process modeling.

Organizational models in use for cellular network processes commonly fall within a few prototypes. These prototypes come with infrastructure of their own. Examples from some of the more prominent prototypes will be illustrated in the latter part of this chapter. First in line will be the *flow network* model. A classic, it provides a major prototype for modeling metabolic and signaling process dynamics as well as applications in revealing related, more permanent network organizational features. Current cellular modeling, however, increasingly turns to models that use graphs as platforms outfitted with other types of infrastructure. With applications at all levels of cellular organization, *Petri nets* (PNs) add unique capabilities to represent ongoing processes and their active control. They are especially valuable in dealing with concurrent processes emphasizing feasibility and potential for misfunction. *Boolean* models, with functional relationships incorporating Boolean logic, draw on network design features and are used in analyzing the workings of various cellular level control machineries, as well as whole-cell properties such as robustness to interventions. Both PNs and Boolean models focus on the organizational aspect but do not represent the quantitative makeup or a realistic time behavior. Kinetic data, however, are increasingly being incorporated into both types of model as they become available. Concluding this chapter will be *motifs*, small cellular networks with well-defined biological activities. Possible blueprints for higher-level modeling, these will illustrate infrastructure that integrates organizational features with quantitative and time factors and how these may be linked to biological function. As a coda, readers will catch a glimpse at how motifs may be identified in large cellular systems based on network structure or activity. Unlike the deterministic models so far, this will feature graphs combined with *statistical* approaches.

Network schematics

What makes for a network in the first place? As is often the case, when dealing with diversity in the real world, the answer depends on context and purpose. Definitions abound. In a most simple view, any three interconnected elements already constitute a network. Graph theory provides a more formal mathematical definition of network, small and large. With cellular function here in mind and in line with system descriptions in previous chapters, networks are defined as communities of functional units in which: (1) members are in functional relationships, (2) interactions within the network or with its surroundings are allowed in all directions—forward, backward, lateral, and cyclical, each individually timed, and (3) the network may consist of components dedicated to diverse tasks.

Networks with schematics such as these are widespread elsewhere in the real world. Modeling networks and their processes in these terms calls for matching abstractions. To begin with, graphs, specially created mathematical objects, will be presented.

Graphs

Modeling networks has its origins in using graphs to represent abstractions of a real world as diverse as bridges between a city and an island (Euler, 18th century, see also below), electric circuits (Kirchhoff, 19th century), and the isomers of organic compounds possible (Cayley, 19th century). These days, graphs and theory are used in a wide range of natural and social sciences, and extensively in technological applications.

The building blocks/elements

Suppose a sample of proteins was obtained from a cellular source. It may consist of receptors, enzymes, structural components, or transporters. Some proteins bind, others do not. Do the binding proteins have a particular organizational structure? What is it? Does it suggest a certain function? To model with a graph (Fig. 8.1) each binding protein will be represented by a circle or a point, often called *vertex* or *node*. Vertices (or vertexes) of the proteins that interact in a functional relationship will be connected by a line segment, an *edge*. Also shown in Fig. 8.1, but not obligatory, vertices and edges may be identified individually by an additional label—a subscript number, letter, or other symbol, for example, v_1, v_2, ..., v_n or e_1, e_2, ..., e_n. Readers suspecting that the protein network in the figure could have been an eukaryotic DNA bound transcription complex would have been correct. In that case v_2 would represent the DNA−TATA box binding protein, v_4 the RNA-polymerase and other vertices the remaining transcription factors.

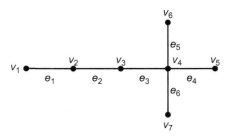

FIGURE 8.1 Simple graph.

Fig. 8.1 depicts a *graph* in which the vertices above represented the cellular components and the edges stand for their functional relationships. Descriptions in this chapter will generally follow the same pattern. Note that there is flexibility here, especially in later approaches where information on functional relationships are also attached to vertices.

Whatever they represent, graphs and their elements are used subject to formal rules. A graph—often denoted G, or else by another uppercase letter—formally consists of a set of vertices, V, *joined* by a set of edges, E. Members from both elements are generally *ordered* in pairs and *alternating*. However, while a vertex may exist by itself, an edge is defined by two *end-points*, most often two vertices and sometimes one and the same (self-edges, see below). A line with only one vertex is not a legitimate edge—it takes two to tango. It is thus possible to create a graph consisting exclusively of vertices, which may remain unconnected, but not of edges only. Using the common meanings of the words, two vertices are *adjacent* if they share an edge on which both are *incident*. Especially in modeling large real-world networks, graphs can include *subgraphs*, separate units of vertices connected by edges internally, but not with other subgraphs, as well as unconnected, solitary vertices.

Network members with class

Cellular networks are heterogeneous. Model vertices then need to be described individually. It can be practical to group together network members that can be distinguished by a particular property in separate classes to begin with. Marking each class by a distinct *color*, known as *coloring*, it is possible for example to distinguish cellular components related to a particular biological function, say members of a control or signaling system, from others that are unrelated. In technological applications, coloring is handy in designing systems so as to prevent interaction among members who might compete, are antagonistic, or provide a dangerous combination. Examples include avoiding college course scheduling conflicts or preventing potential accidents in storing hazardous chemicals. By using colored graphs the design avoids placing vertices with the same or particular color combinations next to each other.

In the same vein a variety of real-world models deal with two class systems aiming to ensure that members do interact, but only outside and not within their own class. These take the form of *bipartite* graphs where each class is represented by a different type of vertex, sometimes depicted by symbols other than circles or dots. By definition, adjacent vertices cannot be of the same class—edges always connect different class members. Bipartite graphs have wide technological as well as social science design applications. From arranging the dinner seating order to maximize the chance of woman—man interactions to avoiding explosive combinations of chemicals in storage. Bipartite graphs can be beneficial in cellular models that depict enzymes and substrates in a pathway by alternating separate vertices. It is a format sometimes favored over the more common versions that use vertices only for enzymes or substrates. Bipartite graphs arise in other cellular process modeling, for example, using PNs, presented later in this chapter. Note also that edges in a graph can be classified by assigning colors along similar lines.

Refining based on functional relationships

The graph in Fig. 8.1, depicting which components interact directly, is an elementary version of an organizational structure in a cellular network. To illustrate what it might take to model many common types of cellular processes, consider a hypothetical scenario. Cells take up glucose by active transport from the exterior. Inside a cell glucose will be used for energy production and other associated processes. These are distributed over a number of cellular compartments. Taking these to be the members of a cellular network and the exchange of products between compartments to be the functional relationships, a model can be constructed as shown in Fig. 8.2. In this scheme, the exterior, vertex v_1, supplies glucose, edge e_1, into the cytoplasm, v_2. To support respiration the cytoplasm provides the mitochondria, v_3, with glucose as well as oxygen as given by edges e_2 and e_3. The mitochondria, in turn, produce ATP, which reaches the cytoplasm for its various uses, represented by e_4. Some ATP is delivered also to the nucleus, v_4, as given by e_5. It is used there for making the glucose transporter mRNA, which then moves for translation into the cytoplasm, e_6. Using oxygen the mitochondria also produce deleterious reactive oxygen species which, as shown, can affect their own activity, e_7. Evidently a model such as this calls for features beyond those of the basic graph in Fig. 8.1.

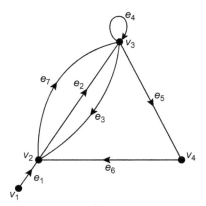

FIGURE 8.2 Directed multigraph.

To begin with a somewhat special device, there is a new type of edge, e_7, representing the potentially self-destructive actions on mitochondria, v_2, of the reactive oxygen species they produce. This edge starts and ends at the same vertex, not at the standard two different vertices and, thus, is a *self-edge* or *loop* (see Fig. 8.2). Used in more special cases this could represent a cell self-stimulated by a paracrine factor, an enzyme activated or inhibited by one of its own products, or a self-regulating transcription factor. The term *loop* is popular in graph literature, but to avoid confusion, it will be reserved here for the different designs of control (open and closed loop and variants).

More generally and extending on the network in Fig. 8.1, cellular systems, especially more complex ones, are often linked by multiple functional relationships. As illustrated, both the mitochondria and the cytoplasm in Fig. 8.2 are connected by *multiple* edges, e_2, e_3, and e_4. Less frequent (and not shown) are several self-edges incident on one and the same vertex as well as multiple vertices combined into single *hyperedges*, convenient in highly populated biological network models (see next).

Cellular processes of interest and the functional relationships between participants generally have a *direction*. In special cases, graphs such as Fig. 8.1 would serve not only to represent structural relationships within networks but also specific processes in which network members constantly associate and dissociate, maintaining an equilibrium. Such *bidirectional processes* are typical in the proteome, but altogether represent a very limited range of the cellular repertoire. More generally, modeling will describe processes that are *unidirectional* such as those in Fig. 8.2— material as well as informational, say, in signaling pathways. As also seen in Fig. 8.2 an edge can be given a direction by adding an *arrowhead* which turns it into an *arc*.

Edge type is a defining feature of the hosting graph. Arcs make for a *directed* graph or *digraph*. With multiple arc connections to individual vertices it is a *directed multigraph*. Edges (undirected) and arcs (directed) may coexist in a *partially directed* or *mixed graph*. The number of arcs leading in or out of a vertex is the *indegree* or *outdegree*, respectively. There is asymmetry between building blocks also here—edges may have a direction, vertices may not. In further handling, ordinary graphs are commonly easier than digraphs and may be adequate for particular process models. Whether or not a digraph is essential, nevertheless, will again depend on the particular problem. Absent self or multiple edges between the same vertices, a graph is *simple*, Fig. 8.1 for example. With multiple edges one has a *multigraph*, with hyperedges a *hypergraph*. Additional vertices and edges create a *supergraph*. A graph with self-edges is a *pseudograph*, slightly lower in mathematical status, but a distinction here of no further concern.

Putting vertices and edges together, there is leeway in displaying graphs on the page. For example, the first graph ever, the Königsberg bridges, appears in textbooks in either of the two equivalent versions shown in

 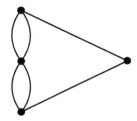

FIGURE 8.3 Konigsberg bridges.

Fig. 8.3. Now, in addition to providing a direct mirror to the real world, a graph is an object with a life of its own. Its rules can be harnessed to serve various practical purposes, as discussed in the following section.

Graphs modified

Graphs can be modified to create or improve models, facilitate handling, or make a theoretical point. Vertices and edges can be *deleted* for various purposes. Sometimes graphs are made smaller and simpler by being *contracted* (i.e., removing edges) thus *merging/consolidating* the incident vertices. On the other hand, *adding* vertices to represent more detail can improve models. The recent trend to upgrade biochemical pathway models by adding to existing vertices for the substrates or for enzymes is important. Inserting or deleting vertices and edges are known as *primary operations.*

Higher level, *secondary operations* are more complex and may be viewed as combinations and repetitions of primary operations. Among others, combining two entire but separate graphs by adding edges to appropriate vertices creates a *join.* Graphs may also be merged into a new supergraphs by *vertex amalgamation/fusion* whereby vertices in separate graphs are made to overlap. Examples of these two operations appear in Fig. 8.4A and B while merging the triangles in Fig. 8.4B would be an example of a contraction generating the graph in Fig. 8.4C. If the graph operations described so far are reminiscent of addition and subtraction in arithmetic there is also a more complicated equivalent for multiplication as can be found in more advanced sources.

The creation of *subgraphs* by reducing a larger parent graph has an important role in modeling as well as manipulating graphs. It is a common way to focus on a salient process or group of functional units without having to deal with an entire network. Technically, subgraphs arise from a parent graph by *deleting* vertices and edges. It is noteworthy that a subgraph is *spanning* when it retains all vertices of the parent graph, although perhaps losing some of the edges. As part of the original, a subgraph in which all vertices are adjacent to one particular vertex is its *neighborhood*, its members the *neighbors.*

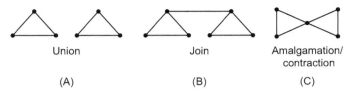

Union	Join	Amalgamation/ contraction
(A)	(B)	(C)

FIGURE 8.4 Graph operations. (A) Union, (B) join, (C) amalgamation/contraction.

Adding value

Aiming for a quantitative model, one might attach a *value* or *weight* to the vertices and/or edges, say, a concentration, an association constant, a reaction or a binding rate. A graph theory definition [1] of networks applies here. Accordingly, networks are directed, mixed, or undirected graphs that are also weighted, that is, edges are assigned real (as opposed to imaginary) number values, including probabilities, or particular mathematical functions that can generate such numbers.

Also of note: (1) Strictly speaking the representation of a graph on the page is a *diagram*, but it is more commonly referred to as a graph. (2) Layout characteristics of vertices and edges, local and global, for historical reasons are often referred to as the graph or network *topology* or *topography*.

Graph properties described so far serve as a general basis for dealing with organizational structure. Specific applications commonly call for additional, more specialized notions with terminology of their own. Materials of general interest will be found in the next two sections. *Walks* will relate to network function. *Trees* will be about structure.

Walks

Connected by a walk

How a given network functions will depend on which components interact. Suppose the metabolism of a microorganism is to be screened for components that would ferment a nutrient into a compound that is hard to synthesize chemically. If the network graph of related metabolism, say, using the enzymes as vertices and their reactions as edges, harbors an appropriate biochemical pathway, the consecutive stages of the process would be described by a (*directed*) *walk*. Technically a walk connecting any two vertices is an undirected or directed continuous sequence of alternating vertices and edges. Either building block alone will specify a walk. The number of edges defines the *length* of a walk. If there is more than one walk connecting two vertices, the shortest is the *distance*.

Various real-world problems call for including all of a network's components exactly *once*, no repetitions allowed. A walk without repeating a vertex or an edge, say, that of the putative fermentation pathway, is referred to here as *path* or *trail*, respectively, regardless of whether it is directed; examples are the walks v_1, v_2, v_3, v_4, or e_1, e_3, e_6, in Fig. 8.1. Walks may be *closed* or *open*, depending on whether or not they start and end at the same vertex. Fig. 8.2 path v_2, v_3, v_4 is closed; Fig. 8.1 paths v_1, v_2, v_3, v_4, or e_1, e_3, e_6, are open. A closed trail with three edges or more is also known as a *cycle* or a *circuit* and is denoted Cn where n is the number of edges (see Fig. 8.2 trail e_2, e_5, e_6)—think of metabolic cycles and feedback loops. It turns out that the presence of cycles can be problematic in working with graphs. There is, consequently, a substantial effort to establish rules that will identify potential for cycles; for example, whenever the degree of every vertex in a graph is at least two, cycles are always present. They are avoided in many a model, but can also be handled.

There is a practical question about a once-only walk that was at the origins of modern graph theory. Would strolling townspeople be able to cross every one of the seven bridges over the river Pregel in the 18th century Prussian city of Königsberg without repetitions and return to the origin? To develop an answer Euler assigned an *edge* to every river Pregel bridge and created a graph of the bridge outlay (Fig. 8.3) in which vertices represented the island and the different river banks. He set out to find out whether the graph would host a closed, now termed *Eulerian*, trail that traverses every edge *once* only. Being closed makes it an Eulerian circuit. Based on a detailed analysis, it turned out that this was impossible, since the vertices were connected by an *odd* number of edges while it might have been possible if their number had been *even*. Note, incidentally, that Eulerian trails have vertex counterparts, in which case they are known as *Hamiltonian* paths.

Readers will recognize that this is essentially an exercise in optimization that aims to minimize the number of stages in a traversal of a system, or its graph. Similar problems, prompted by practical needs, have preoccupied graph theory for a long time. Underlying is, again, the desire to involve *all* of a network's components in the most efficient way. Turning to graphs, this can translate into finding walks that include all of either one of the two graph elements, vertices or edges, but, again, *once* only. Walks, paths, or trails could be open or closed, depending on the model.

Applicable in various other circumstances, a prominent example is the problem of the *traveling salesman*. A salesperson is to leave and return home, visiting every destination in a district *only once* in a district where some destinations are linked directly, but others are not. Choosing among various connections, is there a (closed) walk that minimizes the cost of travelling? This turns out to be a difficult mathematical problem that is usually solved approximately and may call for computer algorithms using *heuristics*, modifications based on experience specific to the particular model.

This type of modeling with open or closed walks has already found related applications in computer executed processes applied to cellular processes. One of these reconstructs nucleic acid sequences from fragments using a related strategy that ensures a unique alignment by generating one, unique walk, an Eulerian trail. Another application seeks to infer protein function from binding data. Graphs offer other optimization tools, seen in the Ford—Fulkerson procedure below.

Network facing interference

Vital to network operation is whether and how it will hold up to harmful interference. As readers well know, disrupting a cellular network, for example by disabling a gene or inhibiting an enzyme, may be lethal or severely damaging. In favorable cases, however, alternative processes can compensate. Naturally there are similar reliability concerns with technological networks—electrical grids, traffic systems, management structures. What would it take to disrupt the normal workings of a network? Are there components that are especially critical? Turning to graph theory this can be seen as an issue of *connectivity* and analyzed in terms of network walks. What follows here will relate primarily to single edge, undirected graphs, but extends also to digraphs and multigraphs.

A graph is *connected* or *disconnected* depending on whether or not it is possible to reach every vertex from another by a walk, that is, an uninterrupted sequence of vertices and/or edges (almost all graphs in this chapter are connected). In case of a network disruption, such a traversal of the model is, by definition, not possible once it has generated a subgraph in which all internal vertices are connected by walks, but those walks in the subgraph are disconnected from the rest of the graph. Such subgraphs are known as graph *components*. One way to determine how reliably a network is internally connected would hinge on whether or not deleting from the vertices or edges of a (connected) graph will increase the number of its components. Deletion of a single, so-called *cut-vertex* may suffice. For example, deleting vertex v_3 from the graph in Fig. 8.1 will qualify as a vertex cut, as it generates two components. One, v_1, v_2 is connected by edge e_1; the second consists of vertices v_4, v_5, v_6, v_7 as connected by edges e_4, e_5, e_6. Note that removing v_3 automatically deletes edges e_2 and e_3. More often, especially given complex large graphs, generating components will require removal of a number of vertices, or a *vertex-cut*. The edge defined counterparts are the *cut-edge* or *bridge* and the *edge-cut*, respectively. An example of a cut-edge or bridge is edge e_1 in Fig. 8.2—when deleted, it generates two components (single vertex v_1 is a legitimate component). Among all cuts whose deletion will disconnect a (large) graph that, with the minimal number of edges, is known as a *minimal edge*-cut or a *minimal-cut set*. These graph operations play a significant role, for example, in characterizing flow networks, the general prototype for network behavior in cellular metabolism and signaling.

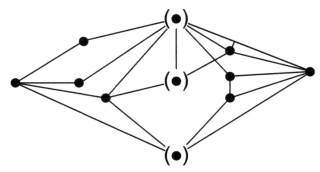

FIGURE 8.5   Separating set.

In quantitative terms there is the minimal number, k, of either graph element needed to disconnect a graph (G), the *vertex-connectivity*, κ_v (G), or the *edge-connectivity*, κ_e (G). Since deleting vertices automatically removes adjacent edges, it usually requires fewer vertex deletions than edge deletions to achieve the same degree of disruption; thus, vertex-connectivity is normally smaller than edge-connectivity. As for terminology, the buyer beware. The term connectivity will have a different meaning in another approach to network structure, described below.

Another index of reliability, not unrelated, is the number of *independent paths* available between two distal vertices. Again, also with an edges counterpart, the goal now is to find the maximum number of *internally* (*vertex* or *edge*)-*disjoint* walks in a network, that is, such that do not share either type of element. According to Menger's fundamental theorem, the maximum number of vertex disjoint, independent paths between two graph vertices equals the minimum number of vertices needed to disconnect them. The same applies for edge and directed graph versions. This translates into finding the minimal number of vertices or edges whose deletion will disconnect the respective graph elements. Called *separating sets*, their number will depend on the complexity of the graph. There is one set separating vertices a and z of an arbitrary graph, the bracketed vertices in Fig. 8.5.

Trees

A botanical hierarchy

Cells harbor networks with a relatively simple organizational structure that can be modeled by a versatile family of graphs called *trees*. Tree models can describe static properties such as patterns of gene expression or the structural interactions in the transcription complex of Fig. 8.1. Tree patterns also have a role in modeling cellular process dynamics. Already familiar as evolutionary and family trees, the prototype is ubiquitous in the real world, notably in computer software for various research applications.

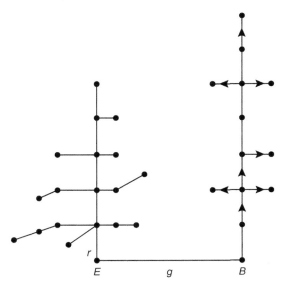

FIGURE 8.6 Tree with rooted subgraph.

A *tree*, depicted in Fig. 8.6, is a connected, simple graph (no multiple and self-edges, or cycles) in which every vertex is reachable by exactly one continuous sequence of edges. In a largely botanical nomenclature, the edges are *branches*. As seen in Fig. 8.6 edges can be undirected, mainly on the LHS, or directed as shown by the arrows on the RHS. With all edges directed, one has a *directed tree* (not shown); Fig. 8.6 is a mixed graph. A vertex at the end of a branch is a *leaf*, however, it can also be a root.

Turning to operations with trees, deleting any one edge from a tree will generate two components; every edge is thus a bridge. Taking it further, for example deleting edge *g* in Fig. 8.6 the parent graph is converted into two subgraphs, *E* on the LHS and *B* on the RHS. Each a tree on its own, together they make a *forest*. Less intuitive and more intriguing is that in graph theory, cutting down either tree will create another forest. Conversely, amalgamating two or more forests can create a single tree. Note also that adding to a tree or a forest no more than a cycle forming edge, here a branch, will disown either from membership in their tree family of graphs. Two types of trees are of special interest, rooted and spanning trees.

Rooted trees

Major cellular processes that are triggered by a single stimulus often branch out—for example, hormone induced intracellular signaling cascades or virus replication. The underlying organizational structure can be depicted by a *rooted tree*, as would family and evolutionary trees. Its *root* is the one and only vertex that has no incoming edge. The root is connected via a directed path to every other vertex of the graph. Fig. 8.6 subgraph E, with vertex r as the root, is an example. By the same token, multiple cellular network processes that feed into a single end product, say, the multiple metabolic reactions producing nucleotides or amino acids, fit well into the rooted tree format as first inverted so that the root becomes an only terminal vertex that represents the final product.

Spanning trees

Connections in real-world networks may be most efficient when they can be described by a so-called *spanning* tree, a graph that spans all vertices and connects them by a minimal number of single edges. Fig. 8.1 is an example. Technological applications are numerous, notably in the design of transportation routes between destinations in larger traffic grids. After assigning realistic weights to the edges, there are computer algorithms that will generate optimal spanning trees. Complex network models are sometimes spanning *forests*, disconnected graphs with internally connected spanning tree components.

Spanning trees can be used, also, to focus on processes associated with a specific component of a network without having to deal with all the others. The King–Altman approach to calculating rate behavior of enzyme species formed with multiple reactants (substrates or effectors—see Chapter 4: Systems/input into output) employs directed spanning trees of a network, selecting only directly contributing reactions.

To conclude on a historical note, it was the listing of all possible spanning trees that allowed graph theory father Cayley to *enumerate* all the possible isomers of given alkanes—carbon hydrogen compounds where carbon atoms were represented as vertices and their bonds as edges.

Structure by numbers

Graphs have quantitative features even before attaching weights. They can be defined by numbers, not only by structural characteristics. The quantitative makeup is often directly relatable to the structure and offers separate tools for working with graphs. Walk connectivity parameters provide a preview.

Graph statistics

To begin with terminology the number of *edges* that are *incident* on a particular vertex is its *degree*, denoted $deg(v)$, with a self-edge being counted twice, and *incidence* having its regular meaning, but also defined technically.

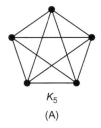 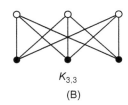

K_5 $K_{3,3}$

(A) (B)

FIGURE 8.7 (A) Complete graph K_5. (B) Bipartite graph $K_{3,3}$.

The total number of vertices in a graph is its *order*, that of the edges its *size*. Maximally connected simple graphs where *each* vertex is connected by an edge to *every* other vertex are frequent in the literature. Called *complete* graphs they are assigned the symbol K_n, where n is the number of vertices. A visually pleasing K_5, order 5, size 10, is shown in Fig. 8.7A and a bipartite $K_{3,3}$ in Fig. 8.7B.

Certain graph properties can be expressed by simple and intuitive statistics. These statistics can provide a quick diagnostic of whether a new graph is constructed properly. They can also have direct applications in characterizing graphs as already used by Euler in deciding the Königsberg bridge traversal problem. There are statistics that are valid, generally; for example, the sum of all vertex degrees is double the number of edges. Vertices with odd degrees should be present in an even number. For a digraph, the sum of *indegrees* and *outdegrees* (i.e., those entering and exiting a vertex) equals the number of edges. There are also numerous statistics typical of particular types of graphs. For instance, in a simple graph there would be at least one pair of vertices with an equal degree and the numbers of vertices, n, edges, m, and components, k, obey the formula $n - k \leq m \leq (n - k)(n - k + 1)/2$. Elsewhere there is a formula for deciding whether a simple graph is in fact connected, which would be the case when of n vertices at least $(n - 1)$ $(n - 2)/2$ edges are connected. Also, an n-vertex graph is always a tree if the number of edges is $n - 1$.

Often used in comparing graphs is the *degree sequence* of the vertices given in nondecreasing or, sometimes nonincreasing order. The degree sequence of the graph in Fig. 8.1, undirected, is thus 1, 1, 1, 1, 2, 2, 4 with a total of 12—an even number twice that of the edges, as required by other rules. Degree sequences can serve to characterize digraphs, such as Fig. 8.2, but employing separate outdegree and indegree sequences. Also typical of a graph is its *degree distribution*—a profile of the fractions of vertices that have the same degree. It can be taken as a measure of the probability of finding a vertex with a certain degree in the entire graph. Returning once again to Fig. 8.1 the degree distribution is 4/7, 2/7, and 1/7, representing vertices with degrees 1, 2, and 4.

Matrix representation

Key graph features are faithfully represented by matrices and tables. It is a format that also makes matrix algebra available for further work up and for a variety of technological applications, for example, in characterizing network connectivity and dynamics. Two matrices, in fact, completely define a graph.

The first, the *adjacency* matrix, A_G, where A and G are common symbols for matrices and graphs, has a row and a column for each vertex. The number of edges between each pair of vertices is entered off-diagonally, that of self-edges; self-edges are entered on the diagonal. A_G is, thus, square and symmetrical. Fig. 8.8 shows the adjacency matrix, A_G, of the graph in Fig. 8.1.

The second is the *incidence* matrix, I_G, where rows again represent the vertices, V_G, but the columns now represent the edges, E_G. When a vertex and an edge are incident the value entered is 1 and 0 otherwise, a self-edge scores 2. The number of vertices and edges not necessarily being equal, I_G is usually not square or symmetrical. Incidence matrices also accommodate vertex directions. As can be seen in Fig. 8.9, the incidence matrix of the directed graph in Fig. 8.2, an outgoing edge is given a $+$ (plus, not shown) and an incoming edge a $-$. Reversing the signs is also legitimate. Note that in the incidence matrix of the same graph, this time undirected, that is unsigned (not shown), the sum of entries in any row equals the degree of the vertex and the sum of the rows equals 2, as expected for a single edge.

However, zero valued entries are information-poor and of limited use. When numerous, as seen especially in Fig. 8.8 and more so with large matrices, they can become a substantial burden in computer operations. More economical and gaining in popularity are tables that list only graph elements that are positively adjacent or incident. Fig. 8.10 is an *incidence table* for the

	v_1	v_2	v_3	v_4	v_5	v_6	v_7
v_1	0	1	0	0	0	0	0
v_2	1	0	0	0	0	0	0
v_3	0	1	0	0	0	0	0
v_4	0	0	1	0	1	1	1
v_5	0	0	0	1	0	0	0
v_6	0	0	0	1	0	0	0
v_7	0	0	0	1	0	0	0

FIGURE 8.8 Adjacency matrix.

	e_1	e_2	e_3	e_4	e_5	e_6	e_7
v_1	1	0	0	0	0	0	0
v_2	-1	1	1	-1	0	-1	0
v_3	0	-1	-1	1	0	0	2
v_4	0	0	0	0	-1	1	0

FIGURE 8.9 Incidence matrix.

$$e_1 \ e_2 \ e_3 \ e_4 \ e_5 \ e_6$$

$$v_1 \ v_2 \ v_3 \ v_4 \ v_4 \ v_4$$
$$v_2 \ v_3 \ v_4 \ v_5 \ v_6 \ v_7$$

FIGURE 8.10 Incidence table.

graph in Fig. 8.1 where each edge has a corresponding column, listing incident vertices below.

Matrices are also used for other organizational structure parameters.

Similarity between graphs

Similarities in organizational structure of two networks may allow for inference from the behavior of a familiar network about another, lesser known network. Criteria for what is similar can be structural as well as quantitative. Graphs of highest similarity are *isomorphic* (equal in features). For practical purposes, this is the case when in both graphs vertices correspond, preserving their adjacencies, vertex-edge incidences, and edge directionalities in digraphs. Other conditions met include both graphs having the same number of vertices and edges as well as the same degree distributions and number of components. As a simple example of an isomorphic pair there are the two versions of the classical bridges graph in Fig. 8.3.

Positively establishing isomorphism between graphs, even when these criteria are met, is often a challenge. Visually, it is an option only with relatively small graphs such as Fig. 8.3. Isomorphism can sometimes be established between more complex graphs by meticulously comparing individual vertices and their adjacencies, but more commonly this will enlist computer programs and will still be problematic. The question arises, then, whether strict isomorphism is absolutely required for networks to be functionally similar. A lesser degree of similarity—for which there is also graph theory, but here out of scope—may be sufficient.

Networks in action

Modeling cellular processes using organizational structure is a diverse methodology that follows different lines of thought. Present-day models have a particular point of view and are geared to specific questions. The rest of this chapter will turn to four prominent prototypes with diverse approaches. These will again be deterministic, currently the predominant format. Concluding this chapter will be a probabilistic coda.

Streaming freely/flow networks

Cellular self-regulating processes in which the collective behavior of pathway members that interact with, or select, neighbors coordinates the overall activity of the system, are well known in metabolic pathways and in signaling cascades (see Chapter 5: Managing processes/control). This is a basic feature of cellular chemistry and ubiquitous in cellular networks, in general. These have an elementary organizational structure which is bypassed in the standard kinetic models (see Chapter 1: Change/differential equations). There is a more general prototype in self-regulating process modeling that extends to entire networks and takes into account their organizational structure. Based on graph theory the technological Ford–Fulkerson *(FF) flow network* model and its methods that originated in the mid-20th century, has, by now, various versions and a range of applications. It is of interest here in its own right and is already informing various approaches in modeling of large metabolic networks. The focus here will be on its basic features.

From source to sink

The Ford–Fulkerson approach to flow networks aims to maximize flow of a commodity in a network at steady state. Mathematically, it is a problem in constrained optimization of *directed flow* in the sense of Chapter 6, Best choices/optimization. Since, to begin with, there was no analytical solution in sight, instead of using rate equations, optimized flows would be identified numerically by computer. This would need a model of the network and an optimization algorithm.

In modeling, say, an urban water-supply system, what should be depicted? The piping system and the directed flow of water in each pipe would be natural choices. Taken together the model becomes a weighted digraph. Assuming that junctions, perhaps pumping stations represented by vertices, can handle any realistic load, the arcs are the critical factor to be specific about, quantitatively, by attaching weights. In a most basic version, shown in Fig. 8.11, flow starts at a *single source* located at a single *s*-vertex, the *source*, and terminates at a single *t*-vertex, the *target* or *sink*. Given there is an overall direction to the flow in the network, locally the model permits

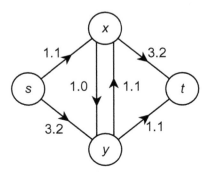

FIGURE 8.11 Flow network.

opposite or sideways flow. In terms of a graph, *forward* arcs may be accompanied by *backward* arcs that go counter to the overall direction.

Flow is specified by attaching to each arc, e, a *capacity*, or *cap(e)*, to make for a *capacitated s-t network*. The capacity stands for the maximal amount of a commodity that can be transferred in a unit time through that part of the network. Capacities of different segments commonly vary; the overall system is discontinuous. In Fig. 8.11 it appears as the first of the two weights attached to each edge. To represent actual flow through the segment, another weighting can be added, the second in Fig. 8.11. The total flow, as measured at vertex *t* equals 3. Current FF models and applications extend well beyond single-source, single-sink directed flow networks. These include models of networks with mixed or undirected flows, as well as networks with multiple sources and multiple targets, connected randomly or along specific routes.

Flow in the network may be subject to two major types of *constraint*. The first constraint is intuitive, namely that actual flow through an arc cannot exceed, but must equal, its flow capacity; that is, it can be *saturating* or less. A minimal flow can be specified, as well. The second constraint, essentially representing conservation of matter in the physical world, is the *conservation of flow* itself. In other words the sum total of flows into all arcs that connect internal vertices must equal that of the flows out. Again, constraints can be imposed on values attached to vertices. When the constraints are satisfied, the flow is *feasible*.

How do networks such as these actually operate? A few points are noteworthy: (1) Flow originates and ends externally. (2) The same process takes place over the entire network. (3) Flow is constantly on, the level of activity of the steady state is at a preset level that depends on the source. Wherever there is input there is also output. (4) Flow takes place in real time with well-defined beginnings and endings. (5) The network is composed of independent segments with discretely varying capacities. Some may be critical to overall flow. Flow is continuous, nevertheless. The overall flow represents the collective behavior of the segments. (6) Especially when networks increase in size, it becomes more likely that what passes through an internal element originated in different locations upstream. Downstream components, however, will generate output regardless of the origin. (7) Flow in high capacity segments of a network may be limited by lower capacity elements upstream and downstream. (8) Whatever control there is, occasionally called *passive*, it rests in the network and is a collective property distributed over its members. While the detail of flow network operations will depend on each case, these more general properties largely overlap with those of cellular counterparts as well.

Optimal flow

That network flow is, in fact, maximal can obviously be a common practical concern. How would maximal flow relate to network structure? As just pointed out, real-world flow networks commonly consist of segments with varying flow-rate—limiting capacities—counterparts of cellular subprocess rate expressions—as shown in Fig. 8.11. Some of the segments may arguably be critical to the flow rate over the entire network and others less so. Identifying critical segments can be based on a notion similar to that used in determining network connectivity. Since flow has a direction, there may be intervening segments whose removal will completely disconnect source and sink. With arcs in a directed graph as counterparts of the edges above, these would make for an *s-t cut*. The cut, in turn, has a capacity of its own. As networks of some complexity may harbor several of these, those with minimal capacity will be a *minimum cut*. The arcs (s, x), (y, x), and (y, t) in Fig. 8.11 provide an example. Intuitively plausible and mathematically proved, there is a *max-flow/min-cut* theorem that states that maximum feasible flow in a network is equal the *capacity* of a minimum cut. The flow value of 3 in Fig. 8.11 is, therefore, also maximal. As are optimization procedures in Chapter 6, Best choices/optimization, FF methods can be used not only to maximize, but also to minimize performance measures, say, cost associated with a process.

A maximizing algorithm

To develop an algorithm that will maximize network flow, the original modeling objective, there is a long line of increasingly sophisticated algorithms designed to cope with large complex networks. Underlying is a common basic scheme, whereby there is initially no flow in the network. A realistic flow is first fed via the source into a path directed toward the target. This could leave unused capacity in, at least, some of the arcs first used and likely also arcs without flow altogether. The algorithm then searches for *flow-* or *(f)-augmenting paths* (meant literally) with unused capacity, including paths that will redirect flow for better use elsewhere or at least free capacity where it might help. Flow is then added accordingly. An iterative procedure, it is repeated until the capacity of connecting arcs is saturated.

Beyond graph theory

All said, the flow network model is indeed a prototype for dealing with self-regulating network processes and their component pathways. Cellular networks, however, host the complexities of diverse, concurrent, interacting, and individually timed processes. Active control and coordination are major factors. Models are needed for longer-term change associated with ageing, development, and, possibly, a varying environment. Models are needed for medical and biotechnological applications as well.

Modeling cellular network processes for these and other purposes turn elsewhere. The rest of this chapter will center on so-called *qualitative* models. Depending on the purpose of modeling and the availability of data, these focus on the course of network processes without necessarily accounting for actual quantities and timings. They become informative when used as platforms for attaching quantitative and time parameters, an increasingly successful trend aiming for comprehensive descriptions of network processes.

Progress by firing squad/petri nets

A versatile approach that uses a graph format and addresses modeling needs of complex networks is available in *PN*s. *PN*s can be used to describe processes that are diverse and take place at the same time, crosstalk, share intermediates, cooperate or interfere, with or without control. Timings of individual processes can be independent. *PN* applications in cell biology so far have been quite diverse—in metabolism, genetic control, signaling, cell division, and differentiation. Introducing *PN*s here will highlight a valuable tool that offers a unique perspective on the workings and modeling of network processes.

Named after their inventor Petri, *PN*s since their inception (1962) have been intimately tied and largely geared toward modeling computer operations. This has resulted in the extensive development of software, a major incentive for turning to *PN* modeling. Extending to a wide range of technological and other real-world networks, *PN*s are used to model networks both locally and globally. From the activity of particular network members in their immediate (organizational) neighborhood to behaviors over larger segments or an entire network, *PN*s are especially popular in handling of concurrent processes. They are often favored in modeling technological processes competing for resources, as well as other causes of conflict.

Process models so far in this book described material change in cellular components, often in the amount or concentration of a particular component. These descriptions employed mathematics and graphs. *PN* process models center on *activity* in a network, that is, on the barebones *events* in which change in the state of the network takes place, no matter what the nature of change is. Using a graph format, *PN*s employ special, customized devices that extend those in graph theory. Motivated by a practical problem solving attitude and backed by detailed mathematical theory, these devices simulate cellular processes, but do not necessarily have actual cellular counterparts. Such models might seem far-fetched, successful when using the ample opportunity they offer to flesh out processes with data and operational rules from the real world. Coverage here will turn first to the activity-centered features of *PN*s; how these can accommodate real-world details will follow.

Building blocks

Overall *PN* usage has evolved in a variety of directions. Early cellular models were generally of the so-called *P/T* (*place/transition*) *net* prototype. Its

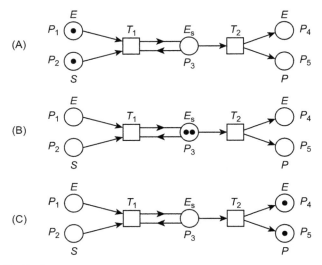

FIGURE 8.12 Petri net.

basic version is often referred to as *standard PN*. The basic building blocks of all versions, however, are universal. Consider, for example, an enzyme reaction pathway. A substrate, *S*, and an enzyme, *E*, form an enzyme substrate complex, *ES*, that will react further to yield the product, *P* and a free enzyme, *E*. A simple scheme of the pathway in *PN* format is depicted in Fig. 8.12. The pathway members, the molecular species that take part in the reaction, are represented by a circle called *place* (*P*), an equivalent of a graph vertex. Turning to the functional relationship, recall that some may be active and others passive. Flow networks make no distinction, but *PN* models represent them separately. The second building block is thus a *transition* (*T*), attached to a neighboring place. Standing for the active component, the transition represents an *action* or an *event* say, the biochemical reactions as they occur. Formally also a vertex, the transition is commonly depicted by a box, as shown, or by a bar.

Places, but not transitions, may harbor for any length of time the third basic element—not used in conventional graphs—a *token* or *mark* (*m*). Tokens are depicted by closed circles or dots, as seen, for example, in those representing the enzyme and the substrate in Fig. 8.12A. Tokens represent the state of the place or a network member. In Fig. 8.12A it means that the process is at the marked stage—the reactants are ready to react. In Fig. 8.12A there is one token per place, but there could be more. The number of tokens in a place is its *marking*, $m(P_i)$. When initiated by a transition, tokens transfer from their original place to an adjacent place changing the state of the donor as well as acceptor places. Tokens can be assigned different meanings, depending on the model, as will be seen below. In Fig. 8.12, for example, token numbers represent the stoichiometry of a reaction, that is, the number of reactants participating.

Still remaining is the familiar *arc*(*A*) (a directed edge), the fourth element. It is here the conduit for the tokens. Arguably a passive aspect of the functional relationship it still represents network structure and the direction of member interactions, sometimes known as the *flow relationship*. Always bridging places with transitions, arcs are often further classified according to whether they represent input (*I*) into or output (*O*) from a transition. For example, the arcs leading from places *E* and *S* into transition *T*1 in Fig. 8.12 are input arcs, those leading from *T*1 to place *ES* and from *T*2 into *P* are output arcs. Since arcs have a direction, a reversible reaction such as *ES* complex formation would be represented by two opposing arcs, as shown in Fig. 8.12A and B. Input and output can define places as well.

Also noteworthy: (1) Seen as graphs *PN*s are directed graphs, bipartite, in that places and transitions are located in tandem vertices that are connected by input and output arcs $P \rightarrow T$ and $T \rightarrow P$. There may be a *capacity* or *weight* attached to places and arcs, respectively indicating the maximum number of tokens allowed to reside or pass through. (2) As other graphs and networks, and in line with much else in this book, *PN*s are linear models that have incidence matrices (see above). There are also matrix representations of *PN* specific properties such as transition, marking, and firing matrices.

In action (local and global)

To set change in motion, the transfer of tokens between places, a transition *fires*. This will happen only when the transition is *enabled*, that is, an adequate number of tokens is available in all directly upstream *pretransition* places. Also required is that tokens can be accommodated downstream—in other words, there is sufficient capacity in receiving, *posttransition* places. These conditions are met before the first firing of the enzyme reaction, as shown in Fig. 8.12B. Note that transition *T*1 requires both the *E* and the *S* tokens to fire and, as shown in Fig. 8.12A, was thus properly enabled. In computer operations, the *PN* origin, information from one stage of a process can be transmitted to multiple destinations—one token turns into several, and vice versa. This can be awkward, although avoidable, in modeling cellular and other real-world processes in which matter needs to be conserved.

How do individual firings behave collectively? Where do network processes take place? What are they like? This calls for collective description of networks where firings take place. Recall that the status of real-world processes, for example in the *SV* model of Chapter 5, Managing processes/control, can be specified by parameters such as the amount of a substance produced or the total amount present, which together make for a *state*. The state of a *PN* is taken to be the distribution of the tokens among all the places and is called a *marking*, *M*. In line with previous process descriptions those described by *PN*s—the sequence and locations of the firings—are given as a succession of stages. These will be represented by the current states of the *PN*, the process being specified by the successions of markings, as seen in Fig. 8.12A–C. Processes that depend on the initial condition, these are given by marking M_0, Fig. 8.12A.

Timings

Modeling the role of organizational features in network dynamics as successions of states—scheduling *when* transitions between states (i.e., firings) should take place—commonly remains unspecified, except in special cases, such as the modeling the self-evident seasons in nature. *Qualitative* models, they arise most often when time behavior data are not available or when they could be inessential. Unlike the simple pathway in Fig. 8.12 dealing with multiple concurrent processes in a larger network, timing can have a direct functional role. It could, for example, make a difference whether it juxtaposes favorable or incompatible stages of different processes. Cellular *PN* models generally specify firing regimes and, thus, become *time(d) PN*s.

Scheduling choices vary. At the elementary end, models activate transitions simply when they are enabled or whenever the process has arrived at a certain stage, as shown in Fig. 8.12A or B. In another option, computer programs include a clock and schedule firings at regular time intervals or at particular time points. Some *PN* programs also allow accounting for time delays between a transition and its outcome—a substantial problem when modeling with equations. Although change in standard *PN*s occurs in steps, that is, inherently discrete, *continuous* time *PN*s are now available as well. *Stochastic PN*s, another option, employ random firings programmed at intervals given by the (statistical) *exponential* distribution which, as readers will recall, was used also for kinetic modeling of stochastic processes in Chapter 7, Chance encounters/random processes.

Feasibility

Turning to *PN* collective features, two will provide examples that relate directly to network behavior and offer tools for network analysis and design. Whether or not, for example, enzymes or signaling proteins in a network are connected is obviously significant to cellular workings. Modeling with *PN*s, there is thus a similar basic question, local and global. Which states will be *reachable* and which will be actually reached by later operations of a network, given an initial state with its specific marking? *Reachability*, as it is known, has practical applications, but, especially in large networks, may be difficult to establish.

At the more local level there are issues of conflict and failure in concurrent processes that can be dealt with by a second type of analysis. Typical in technological networks, cellular biology offers somewhat more benign examples. Storage processes consume precursors, say, the conversion of glucose to starch or glycogen, and, for the time being, stay put. A *PN* model would depict movement of tokens in consecutive metabolic steps ending in the storage material without being able to move on; these tokens are caught in a *trap*. The opposite will occur when the storage material is used up, in which case tokens would be removed without replacement by a *siphon*. Both situations represent *structural deadlock*. Formally these are issues of *liveness*. A transition is *live* when it can fire or, put otherwise, if it enables the next transition. Liveness is sometimes classified further according to whether it extends only to one transition, to a limited sequence within a network or over its entirety. As it were, reachability and liveness are formally related and can be addressed using methods of graph theory as well as those of matrix algebra. Also noteworthy are *PN invariants*, for example, neighborhoods where firing transitions do not change the marking. Determinable from the *PN* incidence matrix, such could be indicative of properties of the steady state of a metabolic network. Invariants also play a role in other more advanced, analyses and design of *PN* models.

Colors, chimeras, hybrids

Current modeling of cellular processes increasingly turns to more recent *PN* versions that are more descriptive, efficient, and include timings. *Colored PNs* (*CPN*, plural *Cp-nets*) already have an extensive technological track record. Cellular modeling emphasizes metabolic (including glycolysis, of course) and signaling pathways as well as genetic control, extending also to higher level cellular systems. Expanding on the *PT* prototype, particularly significant are enhancements of the tokens and the transitions. Enhancing the generic *PT* token above, *CPN* tokens have a *color* each, meaning that each one carries unique information—which process the token represents when there is a choice, perhaps the amount of a participating cellular component, or a mathematical function that can provide a descriptive value. Transition colors may, for example, have attached expressions relating to token transfer, such as preconditions and firing rules. Other advanced features of *CPNs* include places that can harbor a mix of tokens and multiple arcs between places and transitions, possibly in both directions, perhaps dedicated to a particular type of token (i.e., a process). Advanced *Cp-nets* may also have a *hierarchical* arrangement which, in effect, allows a focus on restricted areas of a network structure with the possibility of using different modeling tools.

While *Cp-nets* are deterministic and discrete, continuous, and stochastic timing versions are also available. *PT* and *CPN* versions are formally related. It is thus sometimes advantageous to combine different *PN* methodologies, as in other types of modeling encountered earlier in this book. It could be practical, for example, to start creating a model in a *PT* format and then switch to a *CPN* format, or else have different formats work side-by-side, as a chimera.

A different approach can be taken to modeling, for example, cellular networks that combine genetic control with subsequent biochemical reactions, as in induced enzyme synthesis. In this case one may assign discrete *PN* features to the on/off genetic part and continuous features to the subsequent biochemistry. In another example, a *PN* model of hematopoietic differentiation in human stem cells describes continuously intracellular, primarily signaling related, biochemical processes while choosing whole cell options, in discrete format, of remaining quiescent, dividing, or being committed to differentiation into mature blood cells. Models that combine options, usually only available separately, are in the *hybrid PN* (*HPN*) format and can apply to a wide range of cellular modeling also elsewhere. *Functional hybrid PNs* (*FHPN*) versions, retain the basic *PN* building blocks except that there is a separate set for the discrete and the continuous components which are then put together. Central and unique to *FHPNs* are the self-modifying firing rules associated with the transitions. In this case, the rates of firing tokens can be adjusted to the amount actually present in the supplying preplace, mimicking a realistic (bio)chemical reaction. As modeling with *PNs* constantly evolves, not surprisingly, improved versions emerge that combine (discrete, continuous, stochastic) *HPN* features with those of *CPNs*.

Afterthoughts

(1) Modeling cellular networks with *PNs* obviously have an artificial aspect. Still, *PNs* with attached rate expressions in a sense approximate what would be necessary in setting up a conventional kinetic model, that is, specify when a process would start and stop and how it would run concurrently with others, tasks essentially unmanageable by using only formal mathematics and possibly challenging computer computations. (2) Given their special characteristics, *PNs* share features with other formal approaches, for example, timing arrangements, matrix representations, and logic. Recall that the enzyme reaction was represented in Fig. 8.12 in terms of stoichiometry of the reactants. It could also have been represented as a change in the condition of the reactants, as it might have been in early *PN* versions. The *ES* place in Fig. 8.12B would then carry only one token as would all other places in the figures. This would represent, more directly, that reactants have the options only of being active or staying put—yet another incarnation of the familiar *yes/no*, binary format encountered in previous chapters.

With a little bit of logic/Boolean networks

Can one find principles that will relate organizational structure to network behavior? What might these be? One conceivable way to find out would be to explore organizational structure during ongoing processes by itself, that is, without the quantitative makeup (reaction rates, concentrations, etc.) and timings of the network. These data are often not available to begin with, it may thus be a modeling-only option.

Among various choices, *Boolean networks*, *BN*s, is an approach that employs direct abstractions of the real world. Unlike *PN*s, it is naturally in line with established modeling practice. It has a long history in engineering, notably in modeling electronic circuits and computer operations. *BN* modeling of cellular processes emerged in late-20th century, initiated by Sugita and Kauffman. Originally intended to sidestep nonlinear, continuous mathematics problems in the kinetic modeling of cellular regulatory processes, *BN* modeling would also offer a computer-friendly, discrete mathematics format. Once again qualitative dynamics, *BN* models describe network processes that take place in arbitrary time, as do some of the *PN*s. Obviously partial descriptions, they capture essential features of certain major cellular processes quite well. Models such as these would be especially relevant to cellular networks in which organizational structure, rather than rate parameters, is thought to be decisive in how processes operate.

Since its inception *BN* modeling in systems has continued to focus on cellular control, being applied in prokaryotic and eukaryotic, unicellular and multicellular organisms. Wide-ranging, it extends to models of the control of segment patterning in the embryo of the fruit fly *Drosophila melanogaster*, the life cycles in budding and fission yeasts *Saccharomyces cerevisiae* and *Schizosaccharomyces pombe*, in generic mammalian cells, and elsewhere in the regulation of the protein injecting machinery of the bacterium *Pseudomonas syringae*, biological clocks in the plant *Arabidopsis thaliana*, or the stimulation of apoptosis in murine hepatocytes. In addition, possibly as a main modeling objective, qualitative models can inform on more general features of cellular behavior. Among the more prominent and discussed below is *robustness*—here an assessment of network reliability and stability to perturbation. Vital to the fitness of individual organisms, robustness is thought also to favor success in Darwinian evolution of the species.

Creating *BN* process models will center here on two interrelated features that determine biological function. One is the *design* of the connection patterns represented by graphs and related devices. The other is the *logic* built into the organizational structure that guides the *transition rules*. One way or another these rules draw on the quantitative and abstract representation of human logic, pioneered in the mid-19th century by G. Boole. Hence the name of this approach to modeling. Combining the two components into *BN* models can take different forms.

Some *BN* process models are geared to large-scale network properties and others to the small scale. The *BN* approach in this section will center on networks that affect cellular activity as a whole, reflecting current major applications. Modeling structure and function in small networks, taking into account also quantitative and time behaviors, will be taken up in the next section of this chapter when it turns to cellular motifs. Both types of models make use of infrastructure that is introduced next.

Status and change

Rather than representing a real-world property, *BN* models feature the *status* of network members. The status is always either *ON* or *OFF*, depending on whether a cellular component is present or absent, active, or quiescent. Evidently barebones, these models can be a natural match with cellular processes, notably fast cellular control machinery that, in effect, is always on or off, controlling slow primary process(es). Examples include rapidly activated genes that control slow gene expression processes, and fast-signal transduction in cell-cycle control machinery driving its slowly evolving phases. Entered in graphs, cellular components—a gene or a repressor or members of a signaling cascade—would be represented by vertices whose status would be *ON* or *OFF*. Taking together the current *ON/OFF* status of the entire membership, one has the current *state* of the network. Modeling a process, the initial state is counterpart to the initial conditions described in previous chapters.

In a *BN* model—as with *PN* and *SV* system models in Chapter 4, System/input into output—processes are seen as a sequence of states. To represent a process, *BN* states are successively being *updated* with respect to the *ON/OFF* status of network members. Updating may or may not change the status of a particular network member. This will be determined by *updating* or *transition rules* specific to the model. These rules, however, will be applied depending on the organizational structure of the network. To see how this works, it will be helpful to regard updating of a network member as a process that begins with input from its immediate upstream neighbors and has updated status as an output. Cellular components and their processes often interact with diverse, multiple partners as seen in Figs. 8.1 and 8.2. The upstream input(s) may vary in numbers as well as in its/their possible effect—activating, neutral or deactivating. How a particular input mix will affect the downstream recipient is in *BN* models intimately tied to Boolean or related *logic* and eventually to biological function. What follows is Boolean logic that applies in this section and in modeling motifs in the next section. To begin with basics, it will be convenient to take a brief look at motif logic first.

Logic, qualitative and quantitative

Induced activity in *Escherichia coli* operons is regulated by two upstream factors. As can be seen in Fig. 8.14, activating the *ara*(*binose*)-operon takes both of two regulatory proteins, *CRP* and *AraC*. In more general terms, call

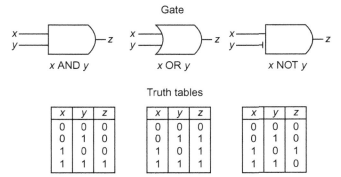

FIGURE 8.13 Logic gates, truth tables.

them factors x and y, as they also appear in the accompanying diagrams. Activation results in making a gene product (GP), z, the measure of activation. Now, change in the status of network component z in the *ara*-operon from *OFF* to *ON* requires both x AND y. On the other hand, the operon synthesizing flagellar motor proteins is activated by protein *FlhDC* (x) and independently by protein *FliA* (y), that is, by x OR y. Then there is the *gal* (*actose*) operon which (in a special strain) is activated by *CRP*, x, but inactivated by protein *GalS*, y. Change in the gene status from *OFF* to *ON* now depends on input that is x NOT y. As it were, *AND*, *OR*, and *NOT* are *logical relations* of which there are others, long-established in analyzing human thought. They are also used in engineering to represent, for example, options for combining digital electronic devices where they are known as *logic gates*. Each of these gates has a symbol of its own. Fig. 8.13 shows the logic gates that correspond to the *AND*, and *OR* relations of the *ara*-operon and the flagellar motor protein operon, above. Also shown is the symbol of the *NOT* relation. Here it is associated with the deactivating input y into the *gal*-operon in what otherwise would be another *AND* relation. By convention it is incorporated into the respective input in the form of a blunt end, ---|, as shown here, or a bubble, ----o, elsewhere.

Given a logic gate or a combination of gates, is there a systematic way to infer the effective input? There is indeed, and it originates in Boole's *logical calculus*. Recall that the model represents whether a cellular component is present or absent or else active or quiescent. This being stated so far in the binary *ON/OFF* status, it could also be given in a *yes/no* format as in previous chapters, or as seen by human logic as *T(rue)* or *F(alse)*. In the logical calculus, *T(rue)* is given the value $T = 1$, and *F(alse)* the value $F = 0$, a common format, for example in byte-level computer coding.

Now, the status effective input will correspond to a so-called *truth value* of the combined inputs, obtained using binary calculus of the particular logical relationships. The truth value is also either true or false and therefore equals 1 or 0, only. Calculating truth values employs common arithmetic operations (addition/subtraction and multiplication). How they are used, however, in this *binary* (1,0) calculus is tailored to human logic and is often the same, but can also differ from usage in conventional algebra. As applied here, and shown in Fig. 8.13, the *AND* relationship, such as that regulating the *ara*-gene, is counterpart to x times y in algebra. Only when both x and y provide input, that is, have the value 1 each, the output also equals 1. The operation *OR*, for example in flagellar motor protein gene control, corresponds to algebraic $x + y$; the output will equal 1 if either factor provides input. The relationship *NOT* negates the actual activation and switches its value where $NOTx = 1 - x$. If the value of either factor would normally have equaled 1, it is now 0, when it is *not* active the output equals 1. Applying this rule to regulating the *gal*-operon, *NOT* negates the actual activation and switches the outcome value to 0. Here, however, it is meaningless when there is no activation to begin with, that is, when $x = 0$. The combined values of all possible inputs can be listed together in a *truth table*, those for the logic gates above are also shown in Fig. 8.13. Complex cellular processes often have components with higher number inputs. In some of these cases models can employ logic gate updating schemes that extend beyond those shown above, but in others one turns elsewhere and here generally beyond scope. The example below employs a simplified formula that bypasses logical gates altogether.

Scheduling updates

Time not being represented as such, as with Petri nets, there are various options *when* and *how* to update *BN*s. Some of the underlying issues, such as how realistic and error free the outcome will be, are similar to those in modeling with *PN*s. There are, however, differences in approach as well as terminology. The simplest to implement and interpret in *BN* modeling is *synchronous* or *in parallel* updating—a deterministic schedule in which all vertices are updated simultaneously. In its way unbiased, it may not be biologically realistic because some cellular processes are fast and others are slow. A synchronous model of the fast machinery controlling the cell cycle may be informative. But modeling the control machinery together with the slower processes in each cellular phase (G_1, S, G_2, and M) could be problematic, potentially creating artificial interaction partners and other significant errors. Turning instead to *asynchronous* updating, schemes come in various versions. In the original and most basic, vertices are updated randomly, one at a time using a formula essentially the same as that in the Gillespie algorithm (see Chapter 7: Chance encounters/random processes). These models have the advantage that there is much already known about the general behavior of these so-called *random Boolen networks*. More advanced versions, for example, allow updating of vertices in groups using schedules that combine a mix of deterministic and stochastic features, including time delays, as well as biological data. Possibly preferable even if not strictly realistic, asynchronous and mixed updating models are generally more difficult to interpret.

An updating process on its way

BN models of large-scale process dynamics in cellular and other real-world networks have certain features in common that can be seen in their graphs. Especially noteworthy are updating pathways that end in particular regions of the network, its *attractors*. They are counterparts of those in *ODE* system dynamics encountered in Chapter 1, Change/differential equations. Attractors being dead ends, further updating will remain within the attractor. Those of cellular network models so far are commonly limited to a few vertices. These can be so-called *limit cycles* among which updates will continue to cycle when updating is continued. Others may be one vertex attractors, *fixed points*. Depending on the network, there may be one attractor or more of various sizes. Particular attractors are often approached from multiple vertices. Pathways that with vertices *ON* reach a particular attractor together make for the attractor's *basin of attraction*.

Attractors are present not only in large network models, they occur also modeling quite small networks. These include a number of logic gates, the *OR* being among them. Known as *canalizing*, at least one input will determine the output of the network regardless of the others, as readers may verify for themselves (see OR truth table Fig. 8.13). Canalizing structures, some larger and not shown, are also likely in larger networks, where they are typically associated with vertices of up to four connections. Such occur also in complex biological networks where they are thought to be significant in the dynamics, overall robustness, and the canalization recognized in developmental processes.

Robustness in a large network

What might *BN* dynamics reveal about the role of organizational structure in higher level cellular behavior? The real-world *BN* modeling examples mentioned earlier represent a range of modeling purposes that used Boolean apparatus such as the above. Here it will be instructive to focus first on what takes place in large-scale modeling, at the same time showing that simplified transition rules can be effective. The detail use of the Boolean rules will be deferred to the next section and the motifs it covers. *Robustness*, a cellular property that is being studied by approaches other than *BN* modeling, is related to stability and reliability properties already encountered in this and previous chapters. Robustness will refer to the propensity of processes to approach and maintain *steady state* both normally and facing internal or external change.

More specifically, consider, for example, the control of the eukaryotic cell cycle. The common cellular machinery is a signaling network that controls the transitions between the different phases of dividing cell life cycle (see above), but itself is self-regulating (see Chapter 5: Managing processes/control) and designated to operate reliably on its own. The apparatus typically includes 10−15 key components that are highly connected. To begin with there is the basic question of how likely it is that its control process will randomly turn to connections that are off its usefully structured course. There are also more particular issues. How will a cellular network respond to modest but realistic change in network components, say, a mutation that increases or decreases its activity, or perhaps eliminates it altogether? Will it withstand changed connections? Would a perturbed process keep its biological function? If so, under what circumstances? Are there network members that are especially critical to robustness? Which? Why? Which network elements are susceptible to external or internal signals? Is developmental or pathological change in a cell associated with change in robustness? Given the complex network and the narrow focus of *BN* model process descriptions, whether a model is valid in the first place is a perennial concern as well.

Addressing questions such as these is the Li model [2] of the cell-cycle control machinery in budding yeast *S. cerevisiae* in which 11 proteins are taken to be key to its operation. Mainly cyclins, cyclin antagonists, protein kinases and, transcription factors, a *BN* model could arguably represent the larger, entire machinery. Underlying the choice of a model is the generally accepted notion that steady states correspond to *BN* attractors. That attractors (i.e., steady states) in fact relate to cell cycle controlling activities in the network would be one indicator of robustness. Other support would rely on the common sense notions that robustness would increase the more the number of pathways that lead to an attractor and the higher is their proportion among the total pathway options available. There could also be more specific arguments, as seen below. There is here, however, a methodological issue typical of modeling large, connected networks. Even random updating could create distinct features, say, groups of vertices with similar status, attractor like in appearance but not directly relevant to the model. Random updating can be simulated and used for comparison. To have biological meaning, that is, to have a genuine structure of its own, a model would have to be significantly different from random. That can be tested by using indicators such as the numbers of attractors, size distributions of basins of attraction, response to change in single network members, and a variety of case specific quantitative relationships.

Turning to the model, employing logical apparatus such as the above in all its detail would be impractical. Instead, the model is *additive*. It is based on a set of relatively simple input relationships that reflect the balance between activating and deactivating interactions of each network member with its upstream neighbors. As seen in Eq. (8.1) there are three updating options that depend on the upstream input shown in a common and new equation format.

$$S_i(t + 1) = \begin{vmatrix} 1, & \sum_j a_{ij}S_j(t) > 0 \\ 0, & \sum_j a_{ij}S_j(t) < 0 \\ S_i(t), & \sum_j a_{ij}S_j(t) = 0. \end{vmatrix} \tag{8.1}$$

The LHS of Eq. (8.1) represents the status of the chosen network component, i, at time $t + 1$, one step after the update at time, t. On the RHS the first column, states that the updated status may take the value 1 (active), 0 (inactive) or $S_i(t)$ (no change). This will depend on the sum $\sum_{ij}$ of the interactions of this component, i, with all its relevant neighbors, j. Each separate input under the sum sign depends, in turn, on the status of the component at the time $S_j(t)$, which could be 1 or 0, and on whether it is activating or deactivating as represented by a coefficient, a_{ij}, that equals 1 or -1, respectively.

To mine for information, the model employs a graph of the consecutive updates in activity status of the network components, representing the latter by the vertices. Drawing conclusions from the model is substantially based on visual information this provides, be it an overview or specifics, notably on the locations of network attractors and basins of attraction. Skipping much detail, the evidence that the model is valid and that budding yeast cell-cycle control is, in fact, robust rests on a number of observations. (1) By indicators such as the above, behavior of the model is structured and not random. (2) Starting the updating process at the natural initiation point, the cyclin Cln_3, the model dynamics reproduce the known sequence of active (*ON*) network members during the life cycle as desired. (3) In an extensive test of robustness in which updating is initiated in all other states of the 2048 possible (2^{11} *ON* or *OFF* states, 11 vertices) all attractors correlate with the network members active at successive stages of the cell cycle. (4) Among the attractors there is one that is highly predominant, attracting almost 90% of the initial states. It corresponds to the major phase of the cell cycle, G_1, and represents the final stage of one complete cycle. Moreover, the pathways that originate on the way to G_1 flow into the sequence of active network members. (5) Changes in makeup of particular network members (activation, inhibition) cause perturbations specific to their associated basins of attraction. A similar, but fission-yeast model indicated, in addition, that control was structured sufficiently to be efficient, but still flexible enough to allow adaptation to change in the network. Observed also in other gene regulatory networks, it is a balance considered favorable to the organism, itself as well as to Darwinian evolution of the species.

All said, there is a constant effort to bring *BN* models closer to real life process behaviors, as there is with *PN*s. Some models employ *thresholds*, say, a particular concentration or rate to be attained as a condition for change in status. Modeling components with multiple inputs, the relative contribution of each is sometimes represented by assigning relative *weights*. When time data on the respective network components are available, network models may actually incorporate an *ODE* for each step. Taking it further there are now protocols for converting Boolean models into full-fledged continuous models that represent fully real life cellular processes. Another type of hybrid model to join the arsenal recently combines *BN* with Petri net features.

With that much about large-scale network behavior, the focus now shifts to the finer detail of organizational structure in action, the small-scale perspective and its modeling. This will center on small teams of network members with specific biological functions whose models integrate Boolean features with kinetic modeling discussed in Chapter 1, Change/differential equations.

Pairing logic with kinetics/motifs

Full-fledged models of biological function that combine network organizational structure with quantitative and time behaviors are a long-term goal, as seen earlier in this chapter. Reckon that combining these components is not trivial. Both the quantitative make up and the timings can determine which network members, in effect, interact and how. In other words quantitative and time factors can affect the workings of the organizational structure. Whether these effects are significant can differ between networks. There are networks that respond to change in rate and timing parameters as one might expect; others are relatively insensitive as pointed out before.

Creating full-fledged models on the scale of cell-cycle control or larger is demanding. Nevertheless, interplay between design, logic, quantitative, and time factors is evident in small, cellular networks where rate and timing information are sufficient and modeling, simpler. Could these small size units also have significant biological function? Some can, and have been extensively characterized in living cells and in the test tube.

Small regulatory devices

Among the more prominent are small, cue-activated networks that regulate gene expression present in prokaryotes and eukaryotes. Members of such networks served to introduce Boolean logic, above. However, how the logic arises and how it eventually works is a matter of the entire control set up. The structure of the networks that regulate the induced transcription of *E. coli ara*, flagellar motor protein, and *gal* gene operons, is depicted in Fig. 8.14A−C. Small networks that are members of large cellular networks, here that of the transcription factors, they are often referred to as network

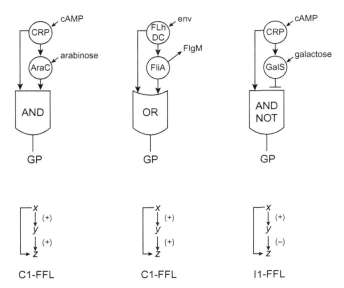

FIGURE 8.14 Motif schematics.

motifs—a term generally used for special function dedicated, recurring cellular patterns in larger structures—for example, the DNA sequences facilitating the initiation of transcription or protein adhesion and phosphorylation sequences.

Control, the biological function of the Fig. 8.14 motifs, is mediated by multistage processes. These can be quite complex but share major features. (1) Small molecular cues bind to the respective regulating proteins enabling them to be active. Binding cAMP, and the sugar arabinose trigger the further downstream transcription of the *ara*-operon, environmental nutrient shortage and among others temperature, pH, and osmotic stresses (*Env*(*ironmental*) in Fig. 8.13) are cues for activating the flagellar motor protein operon and cAMP and the sugars galactose or fucose initiate the activation of the *gal*-operon. (2) Once activated by their cues, one regulatory protein in each of the networks, activating or repressing, acts at the operon level directly. There also can be a second control pathway. In the *ara*, and flagellar motor protein networks, one protein activates the subsequent synthesis of another, also an activating regulating protein. In the *gal* motif, the second protein is a deactivating repressor. It is the cue induced *removal* of this constitutive protein that is activating. Flagellar operon regulator, *FliA*, shown in Fig. 8.13, is activated similarly when bound protein, *FlgM*, dissociates. (3) Regulating proteins that bind to the operon DNA at its promoter activate or inhibit transcription by an RNA polymerase enzyme and eventually the synthesis of GP in Fig. 8.14. For proteins the latter can be used to measure the activation process.

Integrated motif models

Modeling the activities of these *E. coli* motifs has a long history. Keeping the focus on modeling infrastructure, the systematic approach taken by Alon and associates is a valuable source. They also can provide prototypes for larger-scale modeling. Combining the organizational with the kinetic aspects, these models draw on the logic of electrical circuits and represent the dynamics in a biochemical kinetics format. Computer simulations, again, are used extensively, but the detail will not be covered here.

To begin with the organizational component, note that cues activating the motifs act at one end, while control takes place at another. Called *feed forward loops* (*FFLs*), they are well known in technological control and elsewhere in biology as noted in Chapter 5, Managing processes/control. The organizational structure that underlies logic built into these *FFLs* can be seen in Fig. 8.14 diagrams located underneath each of the motif structures. In these diagrams, regulatory proteins x, corresponding to *CRP* and *FlhDC*, provide input for the synthesis of regulatory protein y representing *araC* and *FliA*. Together with proteins *CRP* and *FlhDC* that can act directly, these eventually provide input into the making of the GP z. Synthesis of z is inhibited in the *gal*-operon when there is input from y, the *GalS* protein.

How the organizational structure functions in the synthesis of the GP, z, is linked to still another descriptor, a *sign* attached to each step. As in the additive robustness model of the previous section, it is a *plus* if activating and a *minus* when deactivating. The sign of the two consecutive step pathway on the right in the figures is obtained by multiplying the signs of each step. When the signs in the direct and indirect pathways agree, the motif is *coherent* and when they do not it is *incoherent*. Depending on how member interactions can be activating or deactivating, there are four different types possible of either the coherent or the incoherent arrangements. Since both arabinose control pathways end up with the same sign (plus) as do the flagellar protein pathways, both are *coherent* type 1 FFLs (C1-FFLs). With opposing signs, the control of the galactose operon is an *incoherent* type 1 FFL(I1-FFL). Whether or not the GP, z, is eventually synthesized will depend on the *AND* and *OR* logic gates shown in the network structures above.

However, the ways these networks work also depend on their kinetics, the quantitative behavior over time of the various network components. Activities, being in effect biochemical processes, they can be described by standard (Chapter 1: Change/differential equations) kinetic models of the time-dependent change in the concentrations of the participants. Using *ODEs* that make here for so-called *input functions*, these models are typically Hill functions—expressions closely related to the classical Michaelis−Menten equation, but of no further interest here.

Still, input functions can be simplified for present purposes, incorporating in the process a Boolean component, to become *logic approximations*.

These accommodate a process feature not discussed before. Unlike the cell cycle robustness model, where network activity was taken to be *ON* or *OFF* exclusively, models of the control networks represent a real life continuum between zero and maximal activity. Within this range, there may be activity *thresholds* below or above which a process will not activate or inhibit the next stage. Processes in the motif models here are *ON* only after the responsible cellular component, generally a transcription factor, reaches a threshold concentration.

How are logical approximations applied? Suppose that in the activation process, cue activated factor x further activates the synthesis of factor y. The logic approximation of the activator input function will be given by Eq. (8.2a) and the repressor input function by Eq. (8.2b).

$$\text{(a) } dY/dt = \beta\theta(X > K_a) \qquad \text{(b) } dY/dt = \beta\theta(X < K_r) \tag{8.2}$$

where Y, on the LHS of Eq. (8.1a) and (8.1b) is the concentration of the protein product y, and t is the time. Y is also the output resulting from the input as given on the RHS where β is the maximal rate of the transcribing RNA polymerase, and θ is a Boolean logic component—a binary unit that equals 1 when, as shown in Eq. (8.1a), X, the concentration of protein x, is *above* the threshold concentration for activation, K_a. Representing deactivation, Eq. (8.1b), $\theta = 1$, when X is *below* the threshold, K_r. $\theta = 0$ otherwise in both cases. The thresholds, K_a and K_r, in turn, are taken to equal the concentration X corresponding to half maximal activation or inhibition. In living cells, protein is commonly degraded and sometimes diluted by change in cell size. To represent the concomitant decreases in concentration of the product, Y, both equations will be added to the term, αY, where α is an appropriate coefficient.

A two input process such as the arabinose system and its *AND* logic, for example, will have output that in accordance with the Boolean logical calculus is given by the formula:

$$dZ/dt = \beta_z\theta(X > K_x)\theta(Y > K_y), \tag{8.3}$$

where the RHS again stands for the input and other symbols have the same meaning with K_x and K_y being the activation thresholds in the direct, $x \rightarrow z$ and, $y \rightarrow z$ steps. The flagellar protein *OR* counterpart could take the form

$$dZ/dt = \beta_z\theta(X > K_x \ OR \ Y > K_y) \tag{8.4}$$

and the *lac* operon gate with its repressor would be in the form

$$dZ/dt = \beta_z\theta(X > K_x \ NOT \ Y < K_y) \tag{8.5}$$

In addition to computing input−output relationships such as these, logic approximations also allow computing of other parameters not discussed here, but useful in characterizing these networks, notably response times to addition or withdrawal of the respective cues and steady-state concentrations of network members.

Exercising control

Tested against the real world, motif models integrating organizational structure with kinetics succeed in describing control of gene expression. There are experimental observations in vivo that can be computer simulated effectively, based on integrated models. For one, both types of observation reveal that what in the Fig. 8.14 organizational structure appear as simultaneously acting agents, effectively may involve delays between various steps, a feature due to the kinetics. Moreover, the delays and other time behaviors can differ depending on the built in logic, as it does, for example, between the two coherent ara-, and flagellar operon controls. *Activating* the arabinose *AND* network by its cues, the expression of the gene (i.e., synthesis of component z) is delayed. This is as expected, since both x and y are needed for the activation, where x is available immediately, but y will be present appreciably only after enough x has been produced to activate production of y. Consistent with this interpretation is also that activation of y, but below threshold will not activate the synthesis of z. On the other hand, upon *withdrawal* of the cues, deactivation of operon transcription will start immediately because without cues, proteins x and y, both needed, will soon lose activity. The flagellar protein *OR* gate behaves in reverse: activation is immediate, as x alone can be effective. However, the response to withdrawing the cues is delayed because there is still enough of y present before its natural decay becomes significant.

What biological purpose might be served by the different control designs? Based on the overall features of motif behaviors, physiological context, as well as experience with technological devices there are suggestions. *C1-FFLs*, such as the arabinose and motor protein motifs, are thought to filter off fluctuating stimuli ensuring that the biologically expensive process of operon activation will take place only when absolutely needed and the respective substrates are available. The behavior of the *I1-FFL* controlling the galactose operon makes it a candidate for acting as a rapid-response element, presumably protecting from abrupt withdrawal of the primary nutrient, glucose.

Statistical coda

Interesting as the functional properties of the control motifs may be, questions arise about the overall cellular makeup. Could there be additional *FFLs* of similar design? Perhaps triads or larger motifs with different organizational structure and capabilities? How might they be found? These and related questions concern permanent network properties that are currently being studied intensively and worthy of comment to conclude this chapter. The answers come from permanent and dynamic network features. They have in common being interpreted not deterministically, the predominant approach in this book, but statistically, relating directly to materials in Chapter 7, Chance encounters/random processes. Graphs equipped with statistical parameters provide major infrastructure.

The cellular ecosystem

Searching for motifs in cellular networks—transcriptional or other genetic machinery, protein interaction systems, or metabolic and signaling networks—is an exercise in identifying small groups of components with a particular organizational structure among 1000s or 100s of cellular components. The identity and biological function of most cellular components that can be observed experimentally, say, all proteins, is likely unknown. One has, thus, to assume that cellular network members could interact with any other member in, as yet, unknown ways. Absent definite information, one still might look for groups of components that *probably* could form a motif. As with the large *BN* models in the study of robustness, this would take into consideration that connections between network components could appear randomly. It cannot be ruled out that associations between components that appear random actually have a biological function. Yet, biological cells are highly organized and arguably nonrandom. A systematic search for motifs will, hence, aim first for nonrandom groups of associated network components. What follows will employ quantitative measures of probability and randomness.

Representing by a graph an entire complement of, say, the cellular proteins, there will be high *degree* vertices, that is, vertices with a high *number* of attached edges. Other vertices will be lower degree, perhaps even solitary. The observed frequency of vertices of a particular degree is taken to represent its probability in the entire network. The collective profile of all degree probabilities is the *degree distribution*. Using models, these parameters can be calculated beforehand for a given random network and, presumably, will differ in an organized structure. Classical and frequently used in describing random networks is the *Erdos-Renyi (ER)* model, which begins with a set of vertices with no edges and then computes probabilities of edges being attached at random. The probability of a particular vertex to have k edges, $Pr(k)$ is given in the *ER* model by the (binomial) *Poisson* distribution, encountered in Chapter 7, Chance encounters/random processes. In an undirected graph, $Pr(k) = z^k e^{-z}/k!$, where z is the average degree of the network, namely the sum of all degrees over all vertices divided by total vertex number. The degree distribution in directed random graphs would be given by a related expression.

It turns out that graphs of many and important real-world networks typically display a degree distribution that differs from *ER* random. These networks are *scale-free*. Observed originally in social and technological networks, scale-free in biological cells include metabolic pathways, modules in transcription and genetic control and protein interactions. Instead of being random, the frequencies of vertices of a particular degree and the overall degree distribution in scale-free networks appear to be governed by a power law such that $Pr(k) \sim k^{-\gamma}$, where γ is a constant. The frequencies of high degree vertices are significantly higher in scale-free network models compared to those random. This is consistent with these networks having appreciable internal organization, cells included. To identify particular motifs in this environment there is a spectrum of computer algorithms that are based on statistical considerations and on cues from known network properties.

Guilt by association

If motifs can be identified in the permanent structural features of an overall network, it might be worthwhile also to interrogate network processes in action directly. Data available from high-throughput, microarray profiles of processes in large, cellular networks can be adequate. However, it would be a search for small functional units against a background of possibly 100s or 1000s of components, essentially unknown and random. Nevertheless, methods of *reverse engineering* allow to identify interrelating network members and thus (re)construct, genetic, signaling, and metabolic functions. The approach largely relies on *statistical inference*. Underlying is the notion that when, say, two or more network members have a functional relationship, the probability of their joint appearance in the microarray will be high; otherwise it will tend to random. Depending on the type of biological network and other practical considerations, there are different ways to express and compute these relationships.

Bayesian networks, popular, are directed, cycleless graphs in which the vertices represent network members, say, a gene or a signaling protein, and the arcs represent the interaction of an upstream element affecting a downstream element. The quantitative relationship between the two is the *conditional* probability (see Chapter 7: Chance encounters/random processes) of the downstream element appearing as it depends on that of the presence of an element or more upstream, those immediate and those previous. Using microarray data together with other, already existing knowledge of the network and related theory, this arguably allows to determine the likely parent network that could have originated the observed data—in other words, actual genetic or signaling systems. *Dynamic* Bayesian networks serve to interpret time series profiles, that is, as taken at consecutive times of an ongoing process.

Similar objectives can be achieved based on the familiar notion that probability is linked directly to information and order and inversely related to randomness. In the sense that the more say two genes are likely to appear together, the more, so-called *mutual information* is attributed to each about the other. Dealing with information on network components quantitatively, it is often expressed in terms of logarithmic functions of probabilities of their occurrence. Using logarithmic *joint* probability formulas to represent network member mutual information provides a family of efficient algorithms for identifying network motifs.

Another option to evaluate the strength of association between two network members and presumably the chance of their having a functional relationship are *graphical Gaussian* models. These employ the familiar statistics concepts of *correlation* and the related *covariance*. Readers may recall that the respective *coefficients* will equal zero when the members are statistically independent, but increase when they are related, as will be opportunities to observe them together.

Keep in mind

Biological function in complex cellular networks is intimately linked to the *functional relationships* that determine how and when components and processes interact, and to the overall network *organizational structure* these collectively make. Current cellular process modeling centers on network processes with minimal organizational input using equations of change of previous chapters. When organizational structure becomes significant or is a modeling necessity, additional modeling options include visual representation as well as other abstract tools.

As visual mathematical objects *graphs* offer a general format for modeling organizational structure. Process properties represented can be quantitative and qualitative, dynamic and static, global and local. Rules for constructing, manipulating, and characterizing graphs, as well as interpreting processes, follow *graph theory*. Graphs and theory also provide platforms, language, and points of departure for other network modeling approaches.

Graph building blocks, *vertices* and *edges*, can serve to represent network members and their functional relationships. Vertices and edges can have individualized specifications. When models distinguish between members of particular groups in heterogeneous networks, vertices can be assigned different *colors*. Familiar in technological applications and employed in cellular modeling are two class, *bipartite* graphs that are useful in matching members of different types, or avoiding interactions of similar members.

Depending on the functional relationships between network members, edges can be *undirected* or *directed*. Multiple, and differing interactions between network members can be assigned separate edges between the same vertices. Directed or undirected, possibly mixed, these make for *multigraphs*. For quantitative purposes edges can be assigned *weights*—numerical *values* or *formulas*. Graphs can be manipulated, including adding and subtracting single or multiple building blocks as well as segments and whole graphs. Generating partial, *subgraphs* is common.

Connections within organizational structure relate directly to network performance. Modeling connections can employ *walks*—uninterrupted, directed or undirected sequences of alternating vertices and edges that connect two network vertices. Existence of walks is evidence that cellular pathways are available. Some network processes can be optimized using walks. Network reliability can be evaluated analyzing what walk interruptions—single edge/vertex *cuts* or multiple edge/vertex *cut sets*—will leave a network with unreachable vertices. The number of independent paths a network harbors is another index of reliability.

Connection patterns common in modeling the real world are often described by the graph family of *trees*. Models notably include *rooted* trees that describe processes originating in one vertex; also *spanning* trees, in which all vertices are connected by single edges.

Graph structure can be expressed by quantitative parameters. These relate to the number of elements, distances between vertices and, prominently, *degrees*, the number of edges incident on particular vertices. Arranged by size, *degree sequences* are useful in comparing graphs. Formally, a graph is completely defined by its *adjacency* matrix, entries representing the number of edges between each vertex and by its *incidence* matrix in which entries represent whether or not a particular edge is incident on a particular vertex.

Flow networks are models of process dynamics described by graphs and subject to rules of graph theory. Applicable to self-regulating processes they are prototypes for kinetic modeling of metabolic and signaling pathways. Flow networks receive input and discharge output outside the system. Flow is continuous, generally moving forward, but, locally, also backwards or laterally. It is limited by the capacity of the connecting channels, especially where it is least. Flow is optimizable using iterative computer algorithms that search for unused or unfavorably directed capacity and adjust the flow accordingly.

Network process modeling that goes further, taking into account active control, diverse timings, multiple interacting processes, and features of design and logic, often calls for graphs that represent organizational structure, but process behavior that is described by other theory. Prominent models are *qualitative*, sacrificing time behavior and quantitative detail, and depend on computer implementation. Aiming for full-fledged kinetic modeling, there is an ongoing effort to attach that information where possible.

Petri nets are process simulations that center on relating component *activity* to organizational structure. They are valued especially as a unique tool for dealing with concurrent processes and with general issues of process feasibility, local and global. Using graph elements, processes activity is described by the *transition* powered *firing* that transfers *tokens*, the carriers of change. Active processes are represented by *firing sequences* that can take place depending on formal rules of token supply and demand. *Timings* of *PN* firings may be linked to the state of the network, but often are scheduled, predetermined, or random. Applied to a network, *PNs* can identify long range connections between members that are *reachable*. Local modeling using *liveness* conditions can deal with factors, including *deadlock* between competing processes and *traps*, loci that do not release tokens. Models that approximate real cellular processes are available in advanced versions, such as *colored* and *hybrid PNs*.

BN models describe dynamics as a succession of *states* of network members whose *status* is *ON* or *OFF* and can be assigned values of 1, and 0, respectively. States change, depending on input from upstream network members processed by *logic gates*—mostly activating *AND* and *OR*, possibly modified by deactivating *NOT*. Computing transitions individually is based on *truth tables*. Simplified *additive* transition rules can be useful as seen in a robustness model in which the next state depends on the *balance* between activating and deactivating upstream input. Graphs provide a platform that associates vertices with network members, their status, and transition rules, while edges represent the connections. Common update *timings* are synchronous, random asynchronous, or a mixed schedule.

Common in large-scale, cellular process *BN* models are pathways ending in *attractors*, counterparts of the steady state. Pathways leading to attractors together form a *basin of attraction*. Currently emphasizing complex control and signaling networks, cellular models reproduce their *qualitative* dynamics in a discontinuous format. Ongoing improvements aim for fully *quantitative*, continuous models, that also reproduce time behavior.

Integrated models combining *Boolean logic* with differential equation based *kinetic modeling* provide realistic descriptions of transcriptional control by small, three member networks, *motifs*. Representing quantitative and time behaviors, these models can take into account activity *thresholds*. By design *FFLs*, the in vivo and computer *simulated* motif dynamics depend on the kinetics of their individual steps, activating or inhibiting; also on internal pathways being *coherent* or *incoherent*, and on the logic *AND*, *OR*, and *NOT* gates directly involved in transcription initiation. Identifying motifs within larger networks can be feasible based on *statistical* parameters of network graph structure and experimental process dynamics.

References

[1] Barnes JA, Harary F, et al. Graph theory in network analysis. Soc Networks 1983;5:235−44. Available from: https://deepblue.lib.umich.edu/bitstream/handle/2027.42/25206/0000645.pdf.
[2] Li F, et al. Proc Natl Acad Sci USA 2004;101:4781−6.

Further reading

Alon U. An introduction to systems biology: design principles of biological circuits. Chapman & Hall/CRC; 2007.
Chartrand G. Introductory graph theory. Dover; 1985 (1977).

David R, Alla H. Discrete, continuous, and hybrid petri nets. Springer; 2005.

Erdos P, Renyi A. On the evolution of random graphs. In: Newman M, Barabasi A-L, Watts DJ, editors. The structure and dynamics of networks. Princeton; 2006.

Ford LR, Fulkerson DR. Flows in networks. Princeton; 1962.

Gross J, Yellen J. Graph theory and its applications. CRC; 1998.

Index

Note: Page numbers followed by "*f*" refer to figures.

Printed in the United States
by Baker & Taylor Publisher Services